Studies in Logic

Mathematical Logic and Foundations

Volume 8

A New Approach to Quantum Logic

Volume 1
Proof Theoretical Coherence
Kosta Dosen and Zoran Petric

Volume 2
Model Based Reasoning in Science and Engineering
Lorenzo Magnani, editor

Volume 3
Foundations of the Formal Sciences IV: The History of the Concept of the Formal Sciences
Benedikt Löwe, Volker Peckhaus and Thoralf Räsch, editors

Volume 4
Algebra, Logic, Set Theory. Festschrift für Ulrich Felgner zum 65. Geburtstag
Benedikt Löwe, editor

Volume 5
Incompleteness in the Land of Sets
Melvin Fitting

Volume 6
How to Sell a Contradiction: The Logic and Metaphysics of Inconsistency
Francesco Berto

Volume 7
Fallacies — Selected Papers 1972-1982
John Woods and Douglas Walton, with a Foreword by Dale Jacquette

Volume 8
A New Approach to Quantum Logic
Kurt Engesser, Dov M. Gabbay and Daniel Lehmann

Studies in Logic Series Editor
Dov Gabbay   dov.gabbay@kcl.ac.uk

Mathematical Logic and Foundations editors
         S. Artemov, S. Buss, D. Gabbay, S. Shelah, J. Siekmann, J van Benthem

A New Approach to Quantum Logic

Kurt Engesser
Dov M. Gabbay
Daniel Lehmann

© Individual author and College Publications, 2007. All rights reserved.

ISBN 978-1-904987-53-6
College Publications
Scientific Director: Dov Gabbay
Managing Director: Jane Spurr
Department of Computer Science
King's College London
Strand, London WC2R 2LS, UK

Original cover design by orchid creative   www.orchidcreative.co.uk

---

All rights reserved. No part of this publication may be reproduced, stored in a retrieval system or transmitted, in any form, or by any means, electronic, mechanical, photocopying, recording or otherwise, without prior permission, in writing, from the publisher.

CONTENTS

PREFACE ix

INTRODUCTION xi

CHAPTER 1 A CRASH COURSE IN LOGIC 1
1   Introductory remarks 1
1.1   Syntax of classical propositional logic 2
1.2   A Hilbert style deductive system 3
1.3   Semantics of classical propositional logic 9
1.4   Soundness and completeness 10
1.5   Compactness 11
1.6   Lattices 11
1.7   The Lindenbaum algebra 12
2   Basics of nonmonotonic logic 15
2.1   What is nonmonotonic logic? 15
2.2   Nonmonotonicity in quantum mechanics 16
2.3   Consequence relations and operations 17
2.4   Semantics of nonmonotonic logic 18

CHAPTER 2 SOME HILBERT SPACE THEORY 21
1   The concept of a Hilbert space 21
2   Closed subspaces and projections in Hilbert space 23
3   Orthonormal systems and the Fourier expansion 24
4   More lattice theory 27
5   The lattice of closed subspaces 31
6   Characterising classical Hilbert lattices 33

CHAPTER 3 BASICS OF THE FORMALISM OF QUANTUM MECHANICS   37
1   Some history 37
2   Hermitian operators 39
3   Postulates of quantum mechanics 40
4   Combining systems 42

CHAPTER 4   BIRKHOFF-VON NEUMANN 1936   43
1   Structure of the paper 43
2   Novel logical notions in quantum mechanics. 44
3   Experimental propositions 46
3.1   A propositional calculus for quantum mechanics 47
3.2   The Kochen-Specker and the Schütte tautologies 53

CHAPTER 5   THE DYNAMIC VIEWPOINT: PROPOSITIONS AS
OPERATORS   55
1   Propositions viewed dynamically 55
2   The concept of an M-Algebra 55
3   Motivation and justification 57
3.1   States 57
3.2   Measurements 57
3.3   Illegitimate 58
3.4   Zeros 59
3.5   Idempotence 59
3.6   Preservation 60
3.7   Composition 60
3.8   Interference 61
3.9   Cumulativity 62
3.10   Negation 62
3.11   Separability 62
4   Examples of M-algebras 63
4.1   Examples from classical logic 63
4.2   Orthomodular spaces 64
5   Properties of M-algebras 65
6   Connectives in M-algebras 68
6.1   Connectives for arbitrary measurements 68
6.2   Connectives for commuting measurements 68
7   Commuting measurements 71
8   Separable M-algebras 73

CHAPTER 6   THE LOCAL VIEWPOINT: STATES AS LOGICAL
ENTITIES   75
1   What can logic do about quantum mechanics? 75
2   States as logical entities 78
3   Implication M-algebras 80
4   Conjunction M-algebras 82
5   Strongly separable M-algebras 83
5.1   Basic properties 83

5.2	Encodedness	84
6	No windows theorem: first version	86
7	Limiting case theorem: first version	89
8	The three faces of truth	90

CHAPTER 7 ASPECTS OF QUANTUM REALITY   93

1	The wave particle dualism	94
2	Measurement as an inseparable whole	94
3	Are there "elements of reality"?	96
4	Bohm on wholeness ...	98
5	Experimenting with logic?	100

CHAPTER 8 HOLISTIC LOGICS   105

1	Consequence revision systems	105
1.1	Formal motivation: the Lindenbaum algebra viewed as an operator algebra	105
1.2	The concept of a consequence revision system	106
1.3	Classical logic revisited	114
1.4	The semantics of consequence revision systems: $\mathcal{H}$-Models	115
1.5	$\mathcal{H}$-models in classical logic	117
2	The concept of a holistic logic	118
2.1	Orthogonality, encodedness, dimension	119
2.2	Self-referential soundness and completeness	121
3	No windows theorems: second version	125
3.1	The local no windows theorem	125
3.2	The global no windows theorem	127
4	Limiting case theorem: second version	128

CHAPTER 9 TOWARDS HILBERT SPACE   131

1	Presenting holistic logics	131
1.1	The concept of a Hilbert space logic	131
1.2	The canonical $\mathcal{H}$-Model for a Hilbert space logic	133
1.3	Classical inconsistency in Hilbert space logics	134
2	Symmetry and Hilbert space presentability	135
3	Formal reflections ...	140
3.1	Quantum inference operations	140
3.2	The Birkhoff-von Neumann Extension	142
3.3	The Engesser-Gabbay Extension	142
3.4	The Lehmann Extension	143
3.5	Discussing the extensions	145

CHAPTER 10 FINAL REFLECTIONS — 149

1   The plot of quantum mechanics 149
1.1   Feynman's logical tightrope: the uncertainty principle 151
1.2   How an agent with full introspection can be consistent 153
1.3   The invisible proof operator in classical logic and classical mechanics 159
2   A speculative look 160
2.1   General remarks 160
2.2   The measurement problem in a nutshell 162
2.3   Some more thoughts on measurement 162
2.4   Combining and correlating Hilbert space logics 163
2.5   Representing the measuring instrument 165
2.6   Decomposition, projection, revision in measurement 168
2.7   Is the Hilbert space formalism the whole story? Leggett's macrorealism 171
3   Logical monadology? 173

PREFACE

We are happy to present our book on Quantum Logic to the reader. This research was supported by EPSRC project GR/T24562/01, Quantum Logic and Revision Theory, which enabled the authors to do the research over a period of 18 months during 2004 − 2005 at King's College London. We would like to thank the Department of Computer Science for its support and especially Mrs. Jane Spurr for her superb help with technical and administrative matters. We also had the opportunity during this time to serve the community by organising the Handbook of Quantum Logic and Quantum Structures.

During the preparation of this book Kurt Engesser stayed at Hebrew University twice for a month on the invitation of the Leibniz Center for Research in Computer Science, School of Engineering and Computer Science, Hebrew University, Jerusalem, Israel. This was partially funded by the Alfassa Fund for Research in Artificial Intelligence. He expresses his deep gratitude.

Daniel Lehmann thanks Jean-Marc Lévy-Leblond and Dorit Aharonov for their help.

INTRODUCTION

The main purpose of this monograph is to present the approach to quantum logic developed by the authors in recent years in the coherent form of a book. This approach constitutes a new way of looking at the connection between quantum mechanics and logic.

The message of the book is of interest to a broad audience consisting of logicians, mathematicians, philosophers of science, researchers in Artificial Intelligence and last but not least physicists. These communities, however, strongly differ in their scientific backgrounds. Normally, a physicist has no training in mathematical logic, and a logician is by no means expected to master the Hilbert space formalism of quantum mechanics. This fact constitutes a major problem in any attempt to present the topic of quantum logic in a way accessible to the broad audience to which, in principle, it is of interest. In a journal article for instance it is extremely difficult, if not impossible, to solve this problem. The primary intention of the authors is, apart from giving an extensive and coherent account of their theory, to present their approach in such a way that it is accessible to the heterogeneous audience described.

Therefore, the first chapters serve to provide the reader with the logical and mathematical background that will enable him to understand the subsequent chapters. Chapter 1 presents the prerequisites from logic. In chapter 2 we introduce the concept of a Hilbert space, which constitutes the core structure of the formalism of quantum mechanics. We summarise — for the most part without proof — basic facts of Hilbert space theory which provide the reader with the mathematical equipment necessary for the understanding of the core chapters. Chapter 2 contains a first result of our research, namely a characterisation of classical Hilbert lattices which is of interest from the purely mathematical point of view.

In chapter 3 we give a similar summary of the main facts about the mathematical formalism of quantum mechanics.

Quantum logic has its origin in the famous 1936 now classic paper by Birkhoff and von Neumann entitled "The logic of quantum mechanics" [4]. This paper is still today by far the most widely quoted paper in the field. We devote chapter 4 to a detailed analysis and, in a sense, to a reconstruction of this classic. The reason for this is twofold. The Birkhoff-von

Neumann paper is, today, not easy to read, and it is thus an end in itself to interpret and reconstruct it in modern terminology and highlight its main ideas. Moreover, this chapter serves as a basis for putting in perspective the approach to quantum logic put forward in this book. This approach may, to a considerable extent at least, be viewed as a refinement of the ideas of Birkhoff and von Neumann. The relationship between the two views is on the one hand a 'local-global' relationship. In a sense to be made precise the authors' theory may be viewed as a 'local' version of Birkhoff-von Neumann style quantum logic. On the other hand the refinement consists in our view of propositions. In the approach presented in this book the focus is on viewing propositions as projections in Hilbert space rather than closed subspaces as in the Birkhoff-von Neumann paper. This allows for a *dynamic view of propositions* and has surprisingly far reaching consequences.

The core of the message of the book is contained in chapters 5, 6, 8 and 9. In these chapters we introduce and investigate new concepts which we think can play a fruitful role in quantum logic. Formally, these concepts are abstractions from structures we find in Hilbert space. In chapter 5 we abstract from the lattice of projections of a Hilbert space introducing and studying structures which we call algebras of measurements, M-algebras for short. In coining the term M-algebra for these structures we are aware of the fact that this is a loose way of making use of the term 'algebra'. In the strict sense of Universal Algebra these structures do not qualify as algebras. Logically, a novelty of this approach consists in the way we treat propositions. It is inspired by the analogy between propositions and measurements in physics, in particular quantum measurements. Suppose a physicist performs a measurement of a certain physical quantity A pertaining to a certain physical system. Suppose this system is in a certain state x. The physicist will then in general formulate the result of his measurement as a proposition of the form $A = \mu$, where μ is the value of A measured, and he will then claim this proposition to be a true statement about the system under investigation. In this there is no difference between classical and quantum mechanics. There exists, however, an essential difference between the classical and the quantum case which seems to us of fundamental importance from the logical point of view. Namely, the meanings of the physicist's assertion that $A = \mu$ is true differ in the two cases, classical and quantum. In classical mechanics the proposition $A = \mu$ is a true statement about the physical system in state x. In the quantum case it is a true statement too. It is in the quantum case, however, in general no longer a true statement about the state x but about a certain state y distinct from x, namely about the state of the system 'after measurement'. The reason for this is that quantum measurements generally involve, in contrast to classical measurements, a change of state of the system measured. In chapter 5 we take this dynamic aspect of quantum propositions seriously. It is the source of inspiration for developing a general logical framework in which the *static* notion of truth of a proposition prevailing in traditional logic is replaced by the more general *dynamic* notion of a proposition acting on states. This framework turns out to be a natural generalisation of the traditional static view in which the logical structure of classical and quantum mechanics can be described and their relationship be put in evidence. Classical logic and correspondingly classical mechanics appear as the *static limiting cases* of a *dynamic framework*.

In chapter 6 we introduce what we call the *local viewpoint in quantum logic* as opposed to the global point of view that prevailed in the preceding chapters. In the Birkhoff-von Neumann paper propositions are represented as sets of states. The concept of a state itself is not the focus of attention. This also applies to our framework of M-algebras as developed in chapter 5. In that framework states are primitive notions. In chapter 6 we make the concept of a state itself the focus of investigation. This chapter may be viewed as a logical enquiry into the concept of a physical state. This is pursued further in chapter 8.

Chapter 7 is an interlude addressing primarily those readers who haven't had much contact with quantum mechanics yet. We give a report on some of the well known 'odd' features of the quantum world. Again our intention is twofold. First, the reader can hardly appreciate logical considerations on quantum mechanics without being familiar with the salient physical features of the quantum world. Second, we regard this chapter as a vehicle for conveying the impression to the reader that quantum mechanics touches on fundamental issues, even beyond the realm of physics. It seems that, in contrast to previous physical theories, quantum mechanics raises not just the question what are the laws that govern physical reality but the issue of the very nature of physical reality itself. It is often argued both in the seriously philosophical and the popular scientific literature that the proper understanding of quantum mechanics requires a revision of the view of reality to which we are used from classical physics. It is frequently argued that a main obstacle to the proper understanding of quantum mechanics consists in the fragmenting world view which underlies classical mechanics and, by the way, also classical logic. The intuition all pervading the literature is that quantum mechanics requires a more holistic view of reality than we are used to from classical mechanics. Bohm's classic book "Wholeness and the Implicate Order" [6] is a most profound and eloquent account of this.

In chapter 8 we introduce and study another new concept of crucial importance, namely that of a *holistic logic* and as a special case that of a

Hilbert space logic. Again, the concept of a holistic logic is an abstraction from logical structures we find in Hilbert space as is the concept of an M-algebra.

In chapter 9 entitled "Towards Hilbert Space" we pursue the question whether the concept of a Hilbert space can be characterised in terms of the logical structures studied in the preceding chapters. We present a representation theorem which may be regarded as a positive answer to the above question. This is part of what in the literature on the foundations of quantum mechanics is sometimes called the representation enterprise, a term denoting the project of deriving the formalism of quantum mechanics from certain first principles. In our case these first principles are of a purely logical nature.

In the last chapter, which has a semi-formal and slightly speculative character, we allow ourselves a less formal mode of reflection, even a touch of metaphysics. In particular, we sketch, in a semi-formal and not fully rigorous way, an admittedly speculative treatment of the measurement problem in which the paradoxical nature this problem displays in other approaches is avoided.

As a rough summary one can say that, conceptually, the message of this book rests on two pillars. The first pillar is the *dynamic view of propositions*. We view propositions as acting on states (of the world) and changing them rather than just being true or false in these states. The second pillar is our *logical enquiry into the concept of a physical state*.

Our main mathematical results arise from these two sources.

CHAPTER 1

A CRASH COURSE IN LOGIC

1 Introductory remarks

Logic nowadays means formal logic. Modern logic studies logical systems as *formal systems* based on a precisely defined *formal language*. The concept most central to logic is that of *logical consequence*. Logical consequence is a relation between two statements α and β or, more generally, a relation between a set of statements Σ and a statement α. One may synonymously that say α *is a logical consequence of* Σ or α *follows (logically) from* Σ or α *can be deduced logically from* Σ. Logical deduction is a vital part of our competence as human beings in both everyday and scientific discourse, and it is one of the seminal achievements of modern mathematical logic to have provided the tools for a mathematically rigorous analysis of the intuitive concept of logical consequence.

Given two statements α and β in some (natural or formal) language. What does it mean to say that β is a logical consequence of α? The first idea that may come to mind is to say that β can, in some way, be *proved* from α in the sense that if α is assumed then we can deduce β using certain rules of logical deduction. On this view β *follows from* α means β *is provable from* α. It is obvious that a rigorous analysis must then provide a precise definition of what it means to say *is provable*. In other words, the logician's task then consists in making precise the concept of a proof.

Another natural intuition in approaching the issue of logical consequence is this. We may say that β *follows from* α means something like *whenever α is true, then so is β*. In this case a rigorous treatment requires a 'theory of truth'.

In fact, modern logic reflects these natural intuitions in the way it explicates and studies logical consequence.

Generally, in modern style, logical consequence is specified in a twofold way, namely syntactically and semantically. Its syntactic specification consists in presenting a formal deductive system by a set of logical axioms and a set of rules of deduction. Such a system may look as follows. Given a set Σ of statements and let α be some statement. Then we say that α is a logical consequence of Σ, symbolically $\Sigma \vdash \alpha$, if α can be *proved* from Σ. We then have to say what *proved from* Σ means. Essentially, the idea is this. Assume we have certain purely logical axioms, logical truths so to speak, which can be used freely in any proof. Moreover, consider the statements in Σ as given assumptions that can equally be used freely in the proof. If we can then deduce α from these given statements using the rules of deduction, we say that α can be proved from Σ. A deductive system of this sort is called a *Hilbert style deductive system*. There are various other types of deductive systems such as those introduced by Gentzen or natural deduction type systems introduced by Prawitz. There are also logical systems which put constraints on the use of axioms and assumptions in proofs such as resource logics, in which assumptions are viewed as a sort of resources that can be 'used up' in the course of the proof and can therefore not be used freely in the proof. In this introductory chapter on logic we will present a Hilbert style deductive system that goes back to Hilbert and Bernays. The reader should note that a proof will be a completely formal procedure involving just the formal manipulation of symbols.

The semantic approach to logical consequence invokes the notion of truth. As we said, from the semantic point of view, to say that α is a logical consequence of Σ means that α is true whenever all statements of Σ are true, in symbols $\Sigma \models \alpha$. In this, clearly, two things must be made precise. First we need to make precise what it means to say that a formula is true. Second, we need to make precise what it means to say *whenever it is true*. We will see shortly how this works.

Once we have defined logical consequence in this twofold way, there arises a problem. We then need to study how these two notions of logical consequence, syntactic and semantic, are related. More precisely, we want to prove *soundness* of the logic. This means that we need to show that $\alpha \vdash \beta$ implies $\alpha \models \beta$. This is a natural requirement saying that syntactic consequence implies semantic consequence. If we can prove the other direction too, i.e. that semantic consequence implies syntactic consequence, we say that the logical system is *complete*. Generally a logical system is required to be sound. There are well established logics, however, which are not complete.

In this chapter we present the basics of *classical propositional logic* in the style described. In particular, we will see that classical propositional logic is sound and complete.

1.1 Syntax of classical propositional logic

We start by defining the *language of classical propositional logic*. To those readers who are interested in a presentation of the highest standards we recommend Friedrichsdorf's excellent textbook [19].

1. INTRODUCTORY REMARKS

The language of propositional logic is built up from the following symbols:

- 1) An infinite set of propositional variables
- 2) Symbols for the connectives: $\neg$ (negation) and $\to$ (implication)
- 3) Brackets

We define the set Fml of well formed formulas of the language of propositional logic, formulas for short, inductively by the following clauses:

- Every propositional variable p is a formula.
- If α and β are formulas, so are $\neg \alpha$ and $(\alpha \to \beta)$.

If there is no danger of misunderstandings we omit the brackets. To be more precise, Fml is the smallest set satisfying the above conditions.

Fix a certain variable p and define the symbols $\top$ and $\bot$ as abbreviations for $p \to p$ and $\neg \top$ respectively. Moreover we use the following abbreviations:

- $\alpha \wedge \beta$ for $\neg(\alpha \to \neg \beta)$
- $\alpha \vee \beta$ for $(\neg \alpha \to \beta)$
- $\alpha \leftrightarrow \beta$ for $(\alpha \to \beta) \wedge (\beta \to \alpha)$

1.2 A Hilbert style deductive system

DEFINITION 1.1. A (Hilbert style) deductive system consists of a set of formulas called axioms and a set of rules of inference. Given a deductive system L and a set Σ of formulas. A proof from Σ in L is a sequence of formulas such that each element is either an axiom or an element of Σ or it can be inferred from previous elements using a rule of inference. The elements of Σ are called assumptions. If α is the last element of the sequence, the sequence is called a proof of α from Σ. We say that α is provable from Σ, denoted by $\Sigma \vdash \alpha$, if there exists a proof of α from Σ. If Σ is a set of axioms, we write $\vdash \alpha$ for $\Sigma \vdash \alpha$.

DEFINITION 1.2. $\mathcal{H}$ is a deductive system with four axiom schemes and one rule of inference. More precisely, for any formulas α, β, γ, the following formulas are axioms:

- A1: Axiom scheme 1: $(\alpha \to (\beta \to \alpha))$
- A2: Axiom scheme 2: $((\alpha \to (\beta \to \gamma)) \to ((\alpha \to \beta) \to (\alpha \to \gamma)))$
- A3: Axiom scheme 3: $(\alpha \to (\neg \alpha \to \beta))$

- A4: Axiom scheme 4: $(\alpha \to \beta) \to ((\neg\alpha \to \beta) \to \beta)$

The rule of inference is called (MP for short). For any formulas α, β: if $\vdash \alpha$ and $\vdash \alpha \to \beta$, then $\vdash \beta$:

$$\frac{\vdash \alpha, \vdash \alpha \to \beta}{\beta}$$

The proof of the following important lemma is an exercise in deduction in the above deductive system.

LEMMA 1.3. *For any δ we have $\vdash \delta \to \delta$.*

Proof. Consider axiom scheme $A2$ for $\alpha =: \delta$ and $\beta =: \delta \to \delta$ and $\gamma =: \delta$. Then we have

$$(1)\ \vdash (\delta \to ((\delta \to \delta) \to \delta)) \to ((\delta \to (\delta \to \delta)) \to (\delta \to \delta))$$

Consider axiom scheme $A1$ for $\alpha =: \delta$ and $\beta =: \delta \to \delta$. Then we have

$$(2)\ \vdash (\delta \to ((\delta \to \delta) \to \delta))$$

Modus ponens applied to (1) and (2) gives us

$$(3)\ \vdash (\delta \to (\delta \to \delta)) \to (\delta \to \delta)$$

By $A1$ we have

$$(4)\ \vdash \delta \to (\delta \to \delta)$$

Modus ponens applied to (3) and (4) yields what we want.

$$(5)\ \vdash \delta \to \delta$$

∎

We have the following *derived rules*:

$$R1:$$
$$\frac{\Sigma \vdash \alpha}{\Sigma \vdash \beta \to \alpha}$$

1. INTRODUCTORY REMARKS

$R2$:

$$\frac{\Sigma \vdash \alpha \to (\beta \to \gamma), \Sigma \vdash \alpha \to \beta}{\Sigma \vdash \alpha \to \gamma}$$

$R3$:

$$\frac{\Sigma \vdash \alpha, \Sigma \vdash \neg \alpha}{\Sigma \vdash \beta}$$

$R4$:

$$\frac{\Sigma \vdash \alpha \to \beta, \Sigma \vdash \neg \alpha \to \beta,}{\Sigma \vdash \beta}$$

What does 'derived rule' mean? Consider $R1$. Given a set Σ of assumptions and suppose $\Sigma \vdash \alpha$. Then $R1$ says that $\Sigma \vdash \beta \to \alpha$. We know that there exists a proof of α from Σ: ...α. Considering that $\alpha \to (\beta \to \alpha)$ is an axiom and using modus ponens we see that ...$\alpha, \alpha \to (\beta \to \alpha), \beta \to \alpha$ is a proof of $\beta \to \alpha$ from Σ. As to $R2$ suppose that ...$\alpha \to (\beta \to \gamma)$ and ...$\alpha \to \beta$ are proofs from Σ. Using Axiom scheme 2 and applying modus ponens twice we see that the following sequence is a proof from Σ: ...$\alpha \to (\beta \to \gamma)$...$(\alpha \to \beta), (\alpha \to (\beta \to \gamma)) \to (\alpha \to \beta) \to (\alpha \to \gamma), (\alpha \to \beta) \to (\alpha \to \gamma), \alpha \to \gamma$.

The reader may, in a similar manner, prove that $R3$ and $R4$ are in fact derived rules.

The following lemma states some obvious facts about provability.

LEMMA 1.4.

- (i) If $\Sigma \subset \Omega$ and $\Sigma \vdash \alpha$, then $\Omega \vdash \alpha$

- (ii) Suppose $\Sigma \vdash \delta$ for all $\delta \in \Delta$ and $\Delta \vdash \alpha$. Then $\Sigma \vdash \alpha$

- (iii) Suppose $\Sigma \vdash \alpha$ and $\Sigma \vdash \alpha \to \beta$. Then $\Sigma \vdash \beta$

- (iv) $\Sigma \vdash \alpha$ iff there exists a finite $\Delta \subset \Sigma$ such that $\Delta \vdash \alpha$

DEFINITION 1.5. We call a set of formulas Σ consistent if not $\Sigma \vdash \bot$. We say Σ is maximal consistent if it is consistent and does not admit a proper consistent extension.

LEMMA 1.6. *A set of formulas Σ is consistent iff every finite subset of Σ is consistent.*

Proof. For the direction from left to right assume Σ is consistent and there exists a finite $\Delta \subset \Sigma$ which is inconsistent. This means that $\Delta \vdash \bot$. By lemma 1.4 we then have $\Sigma \vdash \bot$. Thus Σ would be inconsistent contrary to the hypothesis.

For the other direction assume every finite subset of Σ is consistent and Σ is inconsistent. It follows that $\Sigma \vdash \bot$. By lemma 1.4 there exists a finite $\Delta \subset \Sigma$ such that $\Delta \vdash \bot$ contradicting the assumption. ∎

The next lemma is called *Lindenbaum's lemma*.

LEMMA 1.7. *Let Σ be consistent. Then there exists a maximal consistent set Ω such that $\Sigma \subset \Omega$.*

In the sequel we assume the language to be denumerable although, later on in the book, we also want to admit non-denumerable languages. All theorems proved in this chapter also hold for non-denumerable languages. In some of the proofs such as the following we then have to use Zorn's lemma.

Proof. Choose an enumeration $\alpha_0, \alpha_1, \alpha_2...$ of all formulas.

Then define a sequence of sets of formulas $\Sigma_0 \subset \Sigma_1 \subset \Sigma_2...$ recursively as follows: $\Sigma_0 = \Sigma$, $\Sigma_{n+1} =: \Sigma_n \cup \{\alpha_n\}$, if $\Sigma \cup \{\alpha_n\}$ is consistent, otherwise $\Sigma_{n+1} =: \Sigma_n$ Let Ω be the union of all these sets, i.e. $\Omega = \bigcup_{n \in \mathbb{N}} \Sigma_n$

First note that Ω is an extension of Σ by construction.

We claim that Ω is maximal consistent. We first prove that Ω is consistent. To see this, first note that every Σ_n is consistent by construction. Assume Ω is inconsistent. This would by 1.6 mean that a there exists a finite $\Delta \subset \Omega$ that is inconsistent. But we have $\Delta \subset \Sigma_n$ for some n. Thus Σ_n would be inconsistent contrary to the way it was constructed. It follows that Ω is consistent.

We still need to prove that Ω is maximal consistent. Assume it is not maximal consistent. This means that there exists a proper consistent extension Ω^* of Ω which is consistent. Then there exists a formula $\alpha \in \Omega^*$ such that $\alpha \notin \Omega$. We have $\alpha = \alpha_n$ for some n in the above enumeration. Since $\alpha_n \notin \Omega$, we have $\alpha \notin \Sigma_{n+1}$. But this means that $\Sigma_n \cup \{\alpha\}$ is inconsistent and thus Ω^* is inconsistent because $\Sigma_n \cup \{\alpha\} \subset \Omega^*$, which is a contradiction. ∎

The next theorem is called the *deduction theorem*.

THEOREM 1.8. $\Sigma \cup \{\alpha\} \vdash \beta$ *iff* $\Sigma \vdash \alpha \to \beta$

1. INTRODUCTORY REMARKS

Proof. For the direction from right to left assume that there exists a proof of $\alpha \to \beta$ from Σ. Then by modus ponens there exists a proof of β from $\Sigma \cup \{\alpha\}$, which means that $\Sigma \cup \{\alpha\} \vdash \beta$.

The direction from left to right is less obvious. We prove the following. Let $\delta_0, \delta_1, \delta_2, ..., \delta_n = \beta$ be a proof of β from $\Sigma \cup \{\alpha\}$. Then we have $\Sigma \vdash \alpha \to \delta_0, , ..., \Sigma \vdash \alpha \to \delta_n$. Since $\delta_n = \beta$ we are through then.

It is sufficient to show that $\Sigma \vdash \alpha \to \delta_0$ and for $j > 0$ we have $\Sigma \vdash \delta_j$ provided that $\Sigma \vdash \alpha \to \delta_0, ..., \Sigma \vdash \alpha \to \delta_{j-1}$.

For this we consider two cases.

Case 1: $\delta_j \in \Sigma \cup \{\alpha\}$ or δ_j is an axiom. If $\delta_j \in \Sigma$ we have $\Sigma \vdash \delta_j$. Applying the derived rule $R1$ we get $\Sigma \vdash \alpha \to \delta_j$. If $\delta_j = \alpha$ the claim follows by 1.3. This argument also establishes that $\Sigma \vdash \delta_0$.

Case 2: δ_j is obtained by modus ponens from two preceding formulas, i.e. there exists $j_1 \in \{0, ..., j-1\}$ such that δ_j is obtained from δ_{j_1} and $\delta_{j_1} \to \delta_j$ by modus ponens. We have $\Sigma \vdash \alpha \to \delta_{j_1}$ and $\Sigma \vdash \alpha \to (\delta_{j_1} \to \delta_j)$. Then the derived rule R2 gives us $\Sigma \vdash \alpha \to \delta_j$. ∎

LEMMA 1.9. *If not $\Sigma \vdash \alpha$, then $\Sigma \cup \{\neg \alpha\}$ is consistent.*

Proof. We show that $\Sigma \cup \{\neg\alpha\} \vdash \bot$ implies $\Sigma \vdash \alpha$. So let

$$(1) \quad \Sigma \cup \{\neg\alpha\} \vdash \bot$$

We have by lemma 1.3 and the definition of the symbol $\top$

$$(2) \quad \Sigma \cup \{\neg\alpha\} \vdash \top$$

We get by (1) and (2) and the derived rule $R3$ that

$$(3) \quad \Sigma \cup \{\neg\alpha\} \vdash \alpha$$

It follows from (3) and the deduction theorem that

$$(4) \quad \Sigma \vdash \neg\alpha \to \alpha$$

Again by lemma 1.3 we have

$$(5) \quad \Sigma \vdash \alpha \to \alpha$$

The derived rule $R4$ applied to (4) and (5) then gives us

$$\Sigma \vdash \alpha$$

∎

DEFINITION 1.10. Let Σ be consistent. Then call Σ a theory or, synonymously, deductively closed if $\Sigma \vdash \alpha$ implies $\alpha \in \Sigma$. Call Σ complete if for any α we have either $\alpha \in \Sigma$ or $\neg\alpha \in \Sigma$.

THEOREM 1.11. *Let Σ be consistent. Then the following statements are equivalent*

- (i) Ω *is maximal consistent.*
- (ii) Ω *deductively closed, i.e. a theory.*
- (iii) Ω *is complete.*

Proof. We prove that (i) implies (ii). So let Ω be maximal consistent. Consider the set $\Delta := \{\beta \mid \Omega \vdash \beta\}$. We have $\Omega \subset \Delta$. Moreover, Δ is consistent because $\Delta \vdash \bot$ would, by 1.4, imply $\Omega \vdash \bot$ contradicting the consistency of Ω. So Δ is a consistent extension of Ω. Since Ω is maximal consistent, it follows that $\Delta = \Omega$. We have proved that $\Omega \vdash \alpha$ implies $\alpha \in \Omega$ which means that Ω is deductively closed.

We now prove that (ii) implies (iii).

Assume that both α and $\neg\alpha$ are in Ω. Then (i) would give us $\Omega \vdash \alpha$ and $\Omega \vdash \neg\alpha$ and, by the derived rule $R3$ we would have $\Omega \vdash \bot$, which is a contradiction because Ω is assumed to be consistent. It follows that at most one of the formulas α and $\neg\alpha$ are in Ω. Assume now that neither α nor $\neg\alpha$ is in Ω. This would mean, since Ω is maximal consistent, that the sets $\Omega \cup \{\alpha\}$ and $\Omega \cup \{\neg\alpha\}$ are inconsistent, i.e. $\Omega \cup \{\alpha\} \vdash \bot$ and $\Omega \cup \{\neg\alpha\} \vdash \bot$. It would then follow by the deduction theorem that $\Omega \vdash \alpha \to \bot$ and $\Omega \vdash \neg\alpha \to \bot$. By the derived rule $R4$ we would have $\Omega \vdash \bot$, a contradiction.

We still need to prove that (iii) implies (i). Suppose Ω is complete. Assume it is not maximal consistent. Then there exists a consistent extension Ω^* of Ω a formula α such that $\alpha \in \Omega^*$ and $\alpha \notin \Omega$. Since Ω is complete we have $\neg\alpha \in \Omega$ and thus $\neg\alpha \in \Omega^*$. Again using the derived rule $R3$ would give us $\Omega^* \vdash \bot$ contradicting the consistency of Ω^*. It follows that Ω does not admit a proper consistent extension, i.e. that it is maximal consistent. ∎

THEOREM 1.12. *Let Ω be maximal consistent. Then we have $(\alpha \to \beta) \in \Omega$ iff $\alpha \notin \Omega$ or $\beta \in \Omega$.*

Proof. Suppose $(\alpha \to \beta) \in \Omega$ and $\alpha \in \Omega$. We prove that $\beta \in \Omega$. We have $\Omega \vdash \alpha$ and $\Omega \vdash \alpha \to \beta$. By modus ponens we have $\Omega \vdash \beta$ and, since Ω is deductively closed, $\beta \in \Omega$.

For the other direction we need to show that if not $\alpha \in \Omega$, then $(\alpha \to \beta) \in \Omega$, and if $\beta \in \Omega$, then $\alpha \to \beta \in \Omega$.

So assume that not $\alpha \in \Omega$. By 1.11 we have $\neg\alpha \in \Omega$ and thus $\Omega \vdash \neg\alpha$. Using the derived rule $R1$ we then get $\Omega \vdash \alpha \to \neg\alpha$. On the other hand we have by $A3$ $\Omega \vdash \alpha \to (\neg\alpha \to \beta)$. Applying the derived rule $R2$ we get $\Omega \vdash \alpha \to \beta$ and thus, since Ω is deductively closed, $(\alpha \to \beta) \in \Omega$.

Finally assume that $\beta \in \Omega$. By the derived rule $R1$ we have $\Omega \vdash \alpha \to \beta$ and thus $(\alpha \to \beta) \in \Omega$. ■

1.3 Semantics of classical propositional logic

We now approach logical consequence from the semantic point of view. For this we need, as already mentioned, a theory of truth. In the following definition we explain what it means for a formula α to be true under a valuation V.

DEFINITION 1.13. A valuation V is any function assigning a truth value to every formula, i.e. 1 or 0, such that

- $(V)(\neg\alpha) = 1$ iff $V(\alpha) = 0$
- $V(\alpha \to \beta) = 1$ iff $V(\alpha) = 0$ or $V(\beta) = 1$

We say that α is true under V if $V(\alpha) = 1$. If α is true under any valuation V we say that α is a (classical) tautology.

Given a set Σ of formulas and a valuation V such that $V(\alpha) = 1$ for all $\alpha \in \Sigma$ we say that V is a model for Σ.

In the above definition we used the *or* (disjunction) of the English language. In the meta language, i.e. in English, we define the disjunction to be true if at least one disjunct is true.

The reader may think of a valuation V as follows. Given any function $V : Var \to \{0, 1\}$. Then V can be extended in a unique way to a valuation (again denoted) by V. This means that it suffices to specify a valuation by specifying its values for the propositional variables.

In the next lemma we again use a meta connective, namely the *and* (conjunction) of English. In the meta language (English) we define a conjunction to be true iff both conjuncts are true. The following is easily checked.

LEMMA 1.14. *For any valuation V we have*

- $V(\top) = 1$
- $V(\bot) = 0$
- $V(\alpha \wedge \beta) = 1$ *iff* $V(\alpha) = 1$ *and* $V(\beta) = 1$
- $V(\alpha \vee \beta) = 1$ *iff* $V(\alpha) = 1$ *or* $V(\beta) = 1$.

1.4 Soundness and completeness

LEMMA 1.15. *Any maximal consistent Σ has a model. More precisely, there exists a valuation V_Σ such that $V_\Sigma(\alpha) = 1$ iff $\alpha \in \Sigma$.*

Proof. We define V_Σ by its values for the variables as follows. If $p \in \Sigma$ then $V_\Sigma(p) = 1$ otherwise $V_\Sigma(p) = 0$. We prove that V_Σ has the properties above by induction on the construction of formulas.

If α is a variable, the claim holds by the definition of V_Σ.

In the case $\alpha = \neg \beta$ we argue as follows. Suppose $V_\Sigma(\alpha) = 1$. This is the case iff $V_\Sigma(\beta) = 0$. By the induction hypothesis this equivalent to $\beta \notin \Sigma$. Since Σ is complete, this is equivalent to $\alpha \in \Sigma$.

Consider the case $\gamma = \alpha \to \beta$. Suppose $V_\Sigma(\gamma) = 1$. This is the case iff $V_\Sigma(\alpha) = 0$ or $V_\Sigma(\beta) = 1$. But by the induction hypothesis and theorem 1.12 this is equivalent to $\alpha \to \beta \in \Sigma$, i.e. $\gamma \in \Sigma$. ∎

Remark: In the above proof it is only the following properties of Σ that are relevant. First, for any α we have $\alpha \in \Sigma$ or $\neg \alpha \in \Sigma$. Second, for any formula γ of the form $\alpha \to \beta$ we have that $\gamma \in \Sigma$ iff not $\alpha \in \Sigma$ or $\beta \in \Sigma$. It follows that any set Σ having the above properties is maximal consistent.

The following theorem is an immediate consequence of the above lemma.

THEOREM 1.16. *Any consistent set Σ has a model.*

We now give the semantic definition of logical consequence.

DEFINITION 1.17. Given a set Σ of formulas and a formula α. We say that α is a semantic consequence of Σ, in symbols $\Sigma \models \alpha$, if for every model V of Σ we have $V(\alpha) = 1$

It is routine to verify the following. Given any axiom α, then we have for any valuation V that $V(\alpha) = 1$, i.e. every axiom is a tautology. Moreover, given two formulas α and β and a valuation V such that $V(\alpha) = 1$ and $V(\alpha \to \beta) = 1$, then we have $V(\beta) = 1$. Thus any axiom is true under any valuation and modus ponens preserves truth.

The next theorem, which expresses *soundness* of classical propositional logic, is an immediate consequence of the above facts.

THEOREM 1.18. $\Sigma \vdash \alpha$ *implies* $\Sigma \models \alpha$

The following theorem is the *Completeness Theorem* of classical propositional logic. As always in logic, completeness is harder to prove than soundness.

THEOREM 1.19.
  $\Sigma \models \alpha$ *implies* $\Sigma \vdash \alpha$

1. INTRODUCTORY REMARKS

Proof. We show that not $\Sigma \vdash \alpha$ implies that not $\Sigma \models \alpha$. Assume that not $\Sigma \vdash \alpha$. It follows by lemma 1.9 that $\Gamma =: \Sigma \cup \{\neg \alpha\}$ is consistent. By theorem 1.16 Γ has a model, i.e. there exists a valuation V such that $V(\beta) = 1$ for all $\beta \in \Gamma$. In particular we then have $V(\beta) = 1$ for all $\beta \in \Sigma$ and $V(\neg \alpha) = 1$ and thus $V(\alpha) = 0$. But this means that not $\Sigma \models \alpha$ ■

1.5 Compactness

Suppose that $\Sigma \vdash \alpha$. Since any proof of α from Σ involves only finitely many assumptions there exists a finite $\Delta \subset \Sigma$ such that $\Delta \vdash \alpha$.

We therefore have the following theorem, which is known as the *Compactness Theorem*.

THEOREM 1.20. $\Sigma \vdash \alpha$ *(or equivalently $\Sigma \models \alpha$) iff there exists a finite $\Delta \subset \Sigma$ such that $\Delta \models \alpha$ or (equivalently $\Delta \vdash \alpha$)*.

Another version of the compactness theorem is to say that Σ has a model iff every finite $\Delta \subset \Sigma$ has a model. This version follows for instance from the fact that Σ is consistent iff every finite $\Delta \subset \Sigma$ is consistent and that a set of formulas is consistent iff it has a model.

1.6 Lattices

DEFINITION 1.21. A partially ordered set (in short poset) is a pair $\langle L, \leq \rangle$, where L is a non-empty set and $\leq$ is a binary relation satisfying the following conditions

- (i) $A \leq A$ for any $A \in L$ (reflexivity)
- (ii) If $A \leq B$ and $B \leq A$, then $A = B$ (antisymmetry)
- If $A \leq B$ and $B \leq C$, then $A \leq C$ (transitivity)

Call $\leq$ a partial order.

Let $S \subset L$. An upper bound of S is an element $a \in L$ such that $b \leq a$ for all $b \in S$. A least upper bound of S is is an element $a \in L$ such that a is an upper bound and $a \leq b$ for every upper bound of S. Analogously we define the concept of lower bound of S and the concept of a greatest lower bound.

It is readily seen that, if a least upper bound exists for S, then it is unique and analogously for the greatest lower bound.

DEFINITION 1.22. The partially ordered set $\langle L, \leq \rangle$ is called a lattice if for any two elements A and B there exists the least upper bound denoted by $A \vee B$ and the greatest lower bound denoted by $A \wedge B$ and there exists a zero element 0 and a unit element 1, i.e. elements such that for all $A \in L$

we have $0 \leq A$ and $A \leq 1$. The lattice is called complete if any subset of L has a greatest lower bound and a smallest upper bound.

The reader may realise that we denote the least upper bound (greatest lower bound) in a lattice by the same symbol as the propositional connectives of conjunction (disjunction) which should not lead to any confusion. It is readily verified that greatest lower bounds and least upper bounds are uniquely determined if they exist. The same is true for the zero and unit element.

DEFINITION 1.23. We call a lattice L distributive if the following holds for any $A, B, C \in L$

$$A \vee (B \wedge C) = (A \vee B) \wedge (A \vee C)$$

DEFINITION 1.24. Let $\mathcal{L}$ be a lattice. A map

$$A \mapsto A^\perp$$

is called an orthocomplementation and $A^\perp$ the orthocomplement of A if it has the following properties

- (i) $A^{\perp\perp} = A$
- (ii) if $A \leq B$ then $B^\perp \leq A^\perp$
- (iii) $A \wedge A^\perp = 0$
- (iv) $A \vee A^\perp = 1$

We call a $\langle L, \leq, ^\perp \rangle$ an orthocomplemented lattice or simply an ortholattice if $^\perp$ is an orthocomplement of the lattice $\langle \leq \rangle$.

DEFINITION 1.25. A Boolean algebra is an orthocomplemented and distributive lattice.

1.7 The Lindenbaum algebra

We now make the connection between classical logic on the hand and certain algebraic structures on the other. In the case of classical logic it is the concept of a Boolean algebra that constitutes its algebraic counterpart. At this stage we introduce the concept of a Lindenbaum algebra in the traditional manner. We will look at this concept in a new way in chapter 8.

Given a consistent set Σ. We then write $\vdash_\Sigma \alpha$ for $\Sigma \vdash \alpha$. Call two formulas α and β Σ-equivalent, in symbols $\alpha \equiv_\Sigma \beta$, if $\vdash_\Sigma \alpha \leftrightarrow \beta$. We will see that this is in fact an equivalence relation. Denote the equivalence class of α by $[\alpha]$ and denote the set of these equivalence classes by A_Σ. These equivalence

1. INTRODUCTORY REMARKS

classes form a (Boolean) algebra in a natural way. Namely, define $[\alpha] \leq_\Sigma [\beta]$ if $\Sigma \cup \{\alpha\} \vdash \beta$ or equivalently by the deduction theorem $\vdash_\Sigma \alpha \to \beta$. This is well defined, as we will see. Similarly define $[\alpha]^* =: [\neg\alpha]$. Again, this is well defined. The proof of the following theorem is straightforward and is, in other books, sometimes left to the reader as an exercise. We present it here in more detail than is usual because, in this book, we will look at the concept of a Lindenbaum algebra from a more general point of view in chapter 8. The reader can then compare the difference in level of these approaches. The gist of the more general viewpoint of chapter 8 is that we can view the Lindenbaum algebra as an operator algebra in a natural way. This will permit us to generalise the concept of a Lindenbaum algebra and to prove a more general theorem of which the next theorem is a special case.

THEOREM 1.26. $\mathcal{B}(\Sigma) = \langle A_\Sigma, \leq_\Sigma, * \rangle$ *is a Boolean algebra.*

The proof of the above theorem relies on certain well known facts of classical logic, which we will state in the following lemmata. It is routine, and we therefore do not present it in all details. Rather we describe the general procedure of the proof elaborating on just a few typical items. The reader is invited to work out the full proof as an exercise.

It should be noted that theorem 1.26 is a purely syntactic statement and so are the lemmata below. The reader might therefore expect purely syntactic proofs. In fact, this way of proceeding is perfectly feasible. It would, however, as the reader can easily convince himself, be fairly tedious at least compared to the way we will actually proceed.

Rather we will prove the syntactic lemmata below in a more transparent way semantically. This means that by the soundness and completeness theorems all the syntactic statements involved in the lemmata below are equivalent to certain semantic statements. It therefore suffices to prove the semantic equivalents of the lemmata below. All the statements to be proved reduce to proving statements of the form $\vdash_\Sigma \alpha$. By soundness and completeness we know that these statements are equivalent to statements of the form $\Sigma \models \alpha$. In order to prove such a statement we can, by the semantic definition of logical consequence, proceed as follows. Given any valuation V such that $V(\varphi) = 1$ for all $\varphi \in \Sigma$, i.e. V is a model of Σ, then show that $V(\alpha) = 1$.

In the sequel Σ denotes a consistent set of formulas.

LEMMA 1.27.

- (1) $\vdash_\Sigma \alpha \leftrightarrow \alpha$
- (2) If $\vdash_\Sigma \alpha \leftrightarrow \beta$, then $\vdash_\Sigma \beta \leftrightarrow \alpha$
- (3) If $\vdash_\Sigma \alpha \leftrightarrow \beta$ and $\vdash_\Sigma \beta \leftrightarrow \gamma$, then $\vdash_\Sigma \alpha \leftrightarrow \gamma$

Proof. We restrict ourselves to (3). The other cases are proved analogously. Assume that $\Sigma \models \alpha \leftrightarrow \beta$ and $\Sigma \models \beta \leftrightarrow \gamma$. We need to show that $\Sigma \models \alpha \leftrightarrow \gamma$. Let V be any model of Σ. Then we have by the first hypothesis that either $V(\alpha) = V(\beta) = 1$ or $V(\alpha) = V(\beta) = 0$. In the first case we get from the second hypothesis that $V(\beta) = V(\gamma) = 1$. It follows that $V(\alpha \leftrightarrow \beta) = 1$. In the second case we get from the second hypothesis that $V(\alpha) = V(\gamma) = 0$ and thus $V(\alpha \leftrightarrow \gamma) = 1$. ∎

We leave the routine proofs of the following lemmata to the reader as an exercise in the procedure applied above.

LEMMA 1.28.
Suppose that $\vdash_\Sigma \alpha \leftrightarrow \alpha'$ and $\vdash_\Sigma \beta \leftrightarrow \beta'$. Then we have

- (1) $\vdash_\Sigma \alpha \wedge \beta \leftrightarrow \alpha' \wedge \beta'$
- (2) $\vdash_\Sigma \alpha \vee \beta \leftrightarrow \alpha' \vee \beta'$
- (3) $\vdash_\Sigma \neg \alpha \leftrightarrow \neg \alpha'$
- (4) $\vdash_\Sigma (\alpha \to \beta) \leftrightarrow (\alpha' \to \beta')$

LEMMA 1.29.

- (1) $\vdash_\Sigma \alpha \leftrightarrow \alpha$
- (2) If $\vdash_\Sigma \alpha \to \beta$ and $\vdash_\Sigma \beta \to \alpha$, then $\vdash_\Sigma \alpha \leftrightarrow \beta$.
- (3) If $\vdash_\Sigma \alpha \to \beta$ and $\vdash_\Sigma \beta \to \gamma$, then $\vdash_\Sigma \alpha \to \gamma$

LEMMA 1.30.

- (1) $\vdash_\Sigma \alpha \to \alpha \vee \beta$
- (2) $\vdash_\Sigma \beta \to \alpha \vee \beta$
- (3) If $\vdash_\Sigma \alpha \to \gamma$ and $\vdash_\Sigma \beta \to \gamma$ then $\vdash_\Sigma \alpha \vee \beta \to \gamma$.
- (4) $\vdash_\Sigma \alpha \wedge \beta \to \alpha$
- (5) $\vdash_\Sigma \alpha \wedge \beta \to \beta$
- (6) If $\vdash_\Sigma \gamma \to \alpha$ and $\vdash_\Sigma \gamma \to \beta$, then $\vdash_\Sigma \gamma \to (\alpha \wedge \beta)$.
- (6) $\vdash_\Sigma \bot \to \alpha$
- (8) $\vdash_\Sigma \alpha \to \top$

LEMMA 1.31.

- (1) $\vdash_\Sigma \alpha \leftrightarrow \neg\neg\alpha$
- (2) $\vdash_\Sigma (\alpha \wedge \neg\alpha) \leftrightarrow \bot$
- (3) $\vdash_\Sigma (\alpha \vee \neg\alpha) \leftrightarrow \top$

LEMMA 1.32. $\alpha \wedge (\beta \vee \gamma) \leftrightarrow (\alpha \wedge \beta) \vee (\alpha \wedge \gamma)$

We now put the above facts together in order to see that the Lindenbaum algebra is a Boolean algebra.

Proof. We first need to show that $\equiv_\Sigma$ is in fact an equivalence relation. We just verify transitivity. The other conditions are proved analogously. So let $\alpha \equiv_\Sigma \beta$ and $\beta \equiv_\Sigma \gamma$. By the definition of $\equiv_\Sigma$ this says that $\vdash_\Sigma \alpha \leftrightarrow \beta$ and $\vdash_\Sigma \beta \leftrightarrow \gamma$. Lemma 1.27 then gives us $\alpha \equiv_\Sigma \gamma$.

We now need to see that $\leq$ is well defined. For this we must verify that given $\alpha \equiv_\Sigma \alpha'$ and $\beta \equiv_\Sigma \beta'$, $\vdash_\Sigma \alpha \to \beta$ implies $\vdash_\Sigma \alpha' \to \beta'$. But this is lemma 1.28 (4). That $\leq$ is a partial order follows from 1.29.

$[\bot]$ is the smallest element. In fact $\bot \to \alpha$ is a tautology for any α. From the fact $\alpha \to \top$ is a tautology it follows that $[\top]$ is the greatest element.

For given α and β it follows from 1.30 that $[\alpha \wedge \beta]$ is the greatest lower bound of $[\alpha]$ and $[\beta]$. Again from 1.30 we get that $[\alpha \vee \beta]$ is the least upper bound of $[\alpha]$ and $[\beta]$. Finally lemma 1.31 guarantees that $*$ is in fact an orthocomplementation. Distributivity is a consequence of lemma 1.32. ∎

2 Basics of nonmonotonic logic

2.1 What is nonmonotonic logic?

Classical logic is monotonic. Given a set Σ of assumptions and a formula α such that $\Sigma \vdash \alpha$. If we add more assumptions to Σ so as to get Σ^* we will still have $\Sigma^* \vdash \alpha$. 'More information' cannot invalidate inferences drawn on the basis of 'less information'. This is what monotonicity means. In the last decades, logicians studied modes of reasoning that do not have this property. In these so-called *nonmonotonic logics* 'old inferences' may be invalidated by 'new information'. What reason can there be for this phenomenon? One reason is incomplete information. This is for instance the case in common sense reasoning. If we view a common sense reasoner's activity as 'jumping to conclusions' on the basis of certain 'pieces of information', it seems quite natural that certain of his conclusions cannot be maintained in the light of additional information.

Another source of nonmonotonicity is perfect introspection of the (reasoning) agent. Imagine a reasoner, i.e. an agent who can infer propositions from sets of assumptions. Suppose, moreover, this reasoner has an additional ability. Namely assume that whenever he can, in his system of reasoning, infer a certain proposition α from a certain set Σ of assumptions, he can, in the same system, infer the proposition saying "I can infer α" denoted by $I\alpha$ and whenever he cannot infer α from Σ he can infer the proposition "I cannot infer α", i.e. $\neg I\alpha$. The former capability is called positive introspection, the latter is called negative introspection. Assume a consistent agent having both capabilities. We give a (still slightly informal) argument to the effect that such a reasoner cannot be monotonic. So assume he is monotonic. Given a set Σ of assumptions and let α be a proposition the reasoner cannot infer from Σ. By negative introspection he can then infer $\neg I\alpha$. Assume that α is consistent with Σ and can be consistently added to Σ. Then the agent *can* infer α from $\Sigma \cup \{\alpha\}$ and thus by (positive) introspection he can infer $I\alpha$ from the enlarged set of assumptions. Since he is assumed to be monotonic, he can still infer $\neg I\alpha$. But this would mean that he is inconsistent. It follows that he cannot be monotonic.

The branch of nonmonotonic logic that takes its origin in considerations of the above sort is called *autoepistemic logic*, see for instance [3] or [46]. We will come back to this in chapter 10.

2.2 Nonmonotonicity in quantum mechanics

The reader may, at this point, ask the question why we want to consider nonmonotonic logics in our study of quantum logic. The answer is that nonmonotonicity is, from the logical point of view, an essential feature of quantum mechanics. We encounter nonmonotonic (logical) systems in nature so to speak. This has to do with Heisenberg's famous Uncertainty Principle, more generally the uncertainty relations we have in quantum mechanics. What are uncertainty relations? We cannot, at this stage, explain this quantitatively. But, qualitatively, it means the following. Consider, say, the electron of a hydrogen atom and assume a certain physical quantity of this electron, say its (total) energy E, is measured. Through measurement we get a certain value, say μ. Viewing a measurement as a sort of proof we then have 'proved' the proposition $E = \mu$. Now assume we measure the position P of the electron. Again, we get a value, say λ, and we have proved the proposition $P = \lambda$. We are now, used to classical physics and classical logic as we are, inclined to say that we now know the energy and the position of the electron and any subsequent measurement of the energy of the electron could only confirm the proposition "$E = \mu$ and $P = \lambda$". It is an empirical and perhaps slightly surprising fact, however, that this is not the case. A subsequent measurement of the energy of the electron will even with certainty yield a value different from μ. The measurement of position invalidates the result of the measurement of energy. This is essentially what we mean by saying that there exists an uncertainty relation between energy and position. From the point of view of logic this is nonmonotonicity.

2.3 Consequence relations and operations

A consequence relation is a binary relation between formulas satisfying certain intuitive conditions we expect logical consequence to satisfy. We assume the (full) language $\mathcal{L}$ of propositional logic. We state some minimal conditions a consequence relation is supposed to satisfy. The following are the minimal conditions as suggested by Gabbay in [24]. The reader may verify that the classical consequence relation $\vdash$ or equivalently $\models$ defined above in fact satisfies these conditions.

$$\text{Reflexivity}$$
$$\alpha \mid\!\sim \alpha$$

$$\text{Cut}$$
$$\frac{\alpha \wedge \beta \mid\!\sim \gamma,\ \alpha \mid\!\sim \beta}{\alpha \mid\!\sim \gamma}$$

$$\text{Restricted Monotonicity}$$
$$\frac{\alpha \mid\!\sim \beta,\ \alpha \mid\!\sim \gamma}{\alpha \wedge \beta \mid\!\sim \gamma}$$

As observed in the KLM paper, any consequence relation satisfying the above conditions has the following property AND:

$$\frac{\alpha \mid\!\sim \beta,\ \alpha \mid\!\sim \gamma}{\alpha \mid\!\sim \beta \wedge \gamma}$$

For a given consequence relation $\mid\!\sim$ define

$$\alpha \equiv \beta \text{ iff } \alpha \mid\!\sim \beta \text{ and } \beta \mid\!\sim \alpha$$

There is a natural generalisation of the concept of a consequence relation which is called a consequence operation or also an inference operation . Let $A \subset \mathcal{L}$. We may then consider the set of all classical consequences of A and denote if by $Cn(A)$. Cn is thus a function from the power set of $\mathcal{L}$ denoted by $\mathcal{P}(\mathcal{L})$ into $\mathcal{P}(\mathcal{L})$. We call it the consequence operation of classical propositional logic. It has the following properties.

Inclusion $\forall A \subset \mathcal{L},\ A \subset Cn(A)$,

Monotonicity $\forall A, B \subset \mathcal{L}, A \subset B \Rightarrow Cn(A) \subset Cn(B)$,

Idempotence $\forall A \subset \mathcal{L}, \mathcal{C}n(A) = \mathcal{C}n(\mathcal{C}n(A))$,

Note that the above properties do not involve connectives. The consequence operation of classical propositional logic satisfies, moreover, the following conditions involving connectives.

Negation $\forall A \subset \mathcal{L}, \alpha \in \mathcal{L}, \mathcal{C}n(A, \neg \alpha) = \mathcal{L} \Leftrightarrow \alpha \in \mathcal{C}(A)$,

Conjunction $\forall A \subset \mathcal{L}, \alpha, \beta \in \mathcal{L}, \mathcal{C}n(A, \alpha, \beta) = \mathcal{C}n(A, \alpha \wedge \beta)$.

It is a well known result of Tarski that the above conditions characterise the consequence operation of classical propositional logic. Any function $Cn' : \mathcal{P}(\mathcal{L}) \to \mathcal{P}(\mathcal{L})$ satisfying the above conditions coincides with the consequence operation of classical propositional logic.

In nonmonotonic logic we of course do not insist on the requirement of monotonicity.

2.4 Semantics of nonmonotonic logic

How can nonmonotonic consequence relations be presented? As to this problem a breakthrough was achieved in the seminal paper by Kraus-Lehmann-Magidor (KLM) [35]. Namely, it was shown by KLM that certain semantic structures which have become known as KLM models are suitable for this purpose. We present here a slight modification of the original KLM structures introduced in [21].

DEFINITION 1.33.

- A Scott model for Fml is any function $s : Fml \to \{0, 1\}$.

- A $GKLM$ (Generalised Kraus–Lehmann–Magidor) model is a structure of the form $\langle S, <, l \rangle$, where S is a non-empty set, $<$ is a binary relation on S and l is a function associating with each $t \in S$ a set of Scott models $l(t)$. The model is required to satisfy the smoothness condition stated in the next definition.

DEFINITION 1.34. Let $\mathcal{M} = \langle S, <, l \rangle$ be a structure as described in the last definition. Let $t \in S$ and α a formula. Then define the satisfaction relation $t \models \alpha$ as follows:

- $t \models \alpha$ iff for all $s \in l(t)$ we have $s(\alpha) = 1$

- Let $A \subset S$. We say that t is $<$-minimal in A iff for all $t' \in A$ such that $t' < t$ we have $t' = t$. We say that A is smooth iff for every $t \in A$, either t is minimal in A or for some $s \in A$, $s < t$ and s is minimal in A.

2. BASICS OF NONMONOTONIC LOGIC

- Let $[\alpha] = \{t \in S \mid t \models \alpha\}$. We say that $\mathcal{M}$ is smooth iff for all α, $[\alpha]$ is smooth.

- For a smooth model $\mathcal{M}$ we define the consequence relation $\mathrel|\!\sim_\mathcal{M}$ as follows: $\alpha \mathrel|\!\sim_\mathcal{M} \beta$ iff for all t minimal in $[\alpha]$, we have $t \models \beta$.

- Given a consequence relation $\mathrel|\!\sim$ and a smooth model $\mathcal{M}$. We say $\mathcal{M}$ is a model for $\mathrel|\!\sim$ iff $\mathrel|\!\sim = \mathrel|\!\sim_\mathcal{M}$.

We cannot motivate the above definitions at this stage. We leave it by reporting that $GKLM$ models have turned out to be extremely suited for presenting nonmonotonic consequence relations semantically. The concept has a long history taking its origin in investigations on the semantics of conditionals. We will hit on these semantic structures in chapter 9 in a natural way in our study of the consequence relations arising from the formalism of quantum mechanics. They form the natural ingredients of the semantic structures for the logical systems studied in chapter 9.

CHAPTER 2

SOME HILBERT SPACE THEORY

In this chapter we first summarise some material from Hilbert space theory relevant to our purpose. In this we omit the proofs with few exceptions. The reader may find the proofs in most textbooks on Functional Analysis or the books we explicitly quote. We then present a characterisation of certain algebraic structures called classical Hilbert lattices, which play an important role in quantum logic.

1 The concept of a Hilbert space

The core mathematical structure of the formalism of quantum mechanics is that of a Hilbert space.

DEFINITION 2.1. Let H be a vector space over the real or the complex numbers and let $\|..\|: H \to [0, \infty]$ be a function such that

- $\|x\|$ iff $x = 0$
- $\|\lambda x\| = |\lambda| \|x\|$
- $\|x + y\| \le \|x\| + \|y\|$

Then we say that H is a normed space with norm $\|..\|$.

Remark: The above definition also works for the quaternions in place of the reals or the complex numbers. This is essentially also true for the following definitions and theorems. We will refer to the quaternionic case only in section 6 and in chapter 9.

Given a normed space H with norm $\|..\|$, we can define a metric d by $d(x, y) =: \|x - y\|$. Thus a normed (linear) space is a metric space in a natural way. We thus have the topological concepts of continuity, Cauchy-sequence etc. Let us just recall the definition of a Cauchy-sequence. Let M be a metric space with metric d and let $(x_n)_{n \in \mathbb{N}}$ be a sequence in M. We say that $(x_n)_{n \in \mathbb{N}}$ is a Cauchy-sequence if the following holds. For any $\epsilon > 0$ there exists an n_0 such that for any $n, m \ge n_0$ we have $d(x_n, x_m) < \epsilon$.

We call a metric space M *complete* if every Cauchy sequence in M converges. A complete normed space is called a *Banach space*.

21

DEFINITION 2.2. Let H be a vector space over the scalar fields of the real or the complex numbers. A mapping $\langle .\rangle$ from $H \times H$ into the scalar field is called an inner (scalar) product, if the following conditions are satisfied:

- (i) $\langle x_1 + x_2, y\rangle = \langle x_1, y\rangle + \langle x_2, y\rangle$
- (ii) $\langle \lambda x, y\rangle = \lambda \langle x, y\rangle$
- (iii) $\langle x, y\rangle = \overline{\langle y, x\rangle}$
- (iv) $\langle x, x\rangle \geq 0$
- (v) $\langle x, x\rangle = 0$ iff $x = 0$

If $\langle .\rangle$ is an inner product we call $\langle H, \langle .\rangle\rangle$ (or in abuse of notation just H) an inner product space (synonymously a Pre-Hilbert space).

Note that in the above definition it follows from (iii) that $\langle x, x\rangle$ is real so that (iv) makes sense.

We have as a consequence that

- $\langle x, y_1 + y_2\rangle = \langle x, y_1\rangle + \langle x, y_2\rangle$
- $\langle x, \lambda y\rangle = \overline{\lambda}\langle x, y\rangle$

The following is the *Cauchy-Schwarz inequality*. This important inequality will not be explicitly used in the sequel. We mention it because it is basic in the sense that it plays an important part in the proofs of the theorems mentioned (not proved) in the sequel.

Let H be an inner product space. Then

$$|\langle x, y\rangle|^2 \leq \langle x, x\rangle \langle y, y\rangle$$

We have equality iff x and y are linearly dependent.

We call two elements x and y of an inner product space *orthogonal* if $\langle x, y\rangle = 0$. We write $x \perp y$.

An inner product space is a normed space in a natural way, namely we define $\| x \| =: \sqrt{\langle x, x\rangle}$, which is readily verified to be in fact a norm. We call the topology induced by the norm the norm topology. The scalar product is continuous in the norm topology.

In inner product spaces we have the *Pythagorean theorem*: If x and y are orthogonal, then $\|x\|^2 + \|y\|^2 = \|x + y\|^2$.

DEFINITION 2.3. We call an inner product space a Hilbert space if it is complete as a normed space. Equivalently we may say that an inner product space is called a Hilbert space if as a normed space it is a Banach space.

The reader should note that we did not impose any condition on the dimension of a Hilbert space although, historically, the concept arose in connection with infinite-dimensional vector spaces. In fact, the most interesting examples of Hilbert spaces in Functional Analysis are infinite-dimensional function spaces. We thus consider both finite-dimensional Hilbert spaces and infinite-dimensional Hilbert spaces. There are, however, marked differences between the finite-dimensional and the infinite-dimensional case which should be kept in mind. It is for instance true that any finite-dimensional inner product space is a Hilbert space, i.e. complete as a normed space. This does not hold in the infinite-dimensional case. There exist infinite-dimensional inner product spaces that are not complete. Other peculiarities of infinite-dimensional Hilbert spaces concern the role of subspaces and that of bases.

2 Closed subspaces and projections in Hilbert space

Given any vector space H and $S \subset H$. Then we say that S is a subspace of H (in the sense of linear algebra) if $0 \in S$ and if $x, y \in S$ then $\lambda x + \mu y \in S$ for any scalars λ and μ.

Given a finite-dimensional inner product space $\langle H, \langle . \rangle \rangle$ and let S be a subspace of H. If we then restrict the inner product to S and denote its restriction again by $\langle . \rangle$ then $\langle S, \langle . \rangle \rangle$ is again a Hilbert space. We cannot expect this to hold in the infinite-dimensional case because we cannot take for granted that $\langle S, \langle . \rangle \rangle$ is complete. If, however, we require S to be closed then $\langle S, \langle . \rangle \rangle$ is in fact a Hilbert space. It is the *closed subspaces* that in the infinite-dimensional case play, essentially, the role of subspaces in the finite-dimensional case. Let us take a closer look at closed subspaces of a Hilbert space. To be precise, we call a subset S of a Hilbert space a closed subspace if S is a subspace in the sense of linear algebra and if it is a closed set in the norm topology.

For a given subset S of a Hilbert space we denote its closure, i.e. the smallest closed set containing S, by $\bar{S}$. It is easily seen that for any subspace S its closure $\bar{S}$ is a closed subspace. For two (not necessarily closed) subspaces A and B we denote by $A + B$ the smallest (not necessarily closed) subspace containing A and B. For two closed subspaces A and B denote by $A \vee B$ the smallest closed subspace containing A and B. Clearly we have $A + B \subset A \vee B$. We will see that this may be a proper inclusion. Call two subspaces A and B orthogonal if for any $x \in A$ and $y \in B$ we have $x \perp y$.

For any subspace A define its orthogonal complement by

$$A^\perp =: \{x \in H \mid (\forall y \in A) x \perp y\}$$

$A^\perp$ is again a subspace. If A is closed, so is $A^\perp$. Moreover, A is a closed subspace iff $A = A^{\perp\perp}$ The following theorem is well known.

THEOREM 2.4.

Let A be a closed subspace of the Hilbert space H and $x \in H$. Then we have

$$H = A \oplus A^{\perp},$$

i.e. any $x \in H$ has a unique decomposition $x = y + z$ with $y \in A$ and $z \in A^{\perp}$.

Given a closed subspace A and $x \in H$. Let $x = y + z$ the unique decomposition of x as above. Then we call y the projection of x onto A denoted by $P_A(y)$. We refer to the mapping P_A just as A context permitting. We call these mappings *projections*. A mapping $T : H \to H$ is thus called a projection if there exists a closed subspace A such that $T = P_A$. Any projection P_A is bounded, linear and *idempotent*, i.e. we have $P_A^2 = P_A$. We have that $A = \{x \in H \mid P_A x = x\}$ and $P_A = P_B$ iff $A = B$.

THEOREM 2.5 (Projection Theorem). *Let H be a Hilbert Space, $x \in H$ and A be a closed subspace of H. Then there exists a unique $y \in A$ such that $d(x, y) = \inf_{z \in A} d(x, z)$ and we have $P_A x = y$.*

3 Orthonormal systems and the Fourier expansion

Recall from linear algebra that any vector space and thus any Hilbert space — finite-dimensional or not — admits a basis. A basis in the sense of linear algebra is a family of linearly independent vectors that spans the whole space. Every vector has then a unique representation as a linear combination of basis vectors. In the case of infinite-dimensional Hilbert spaces it is not the bases in the sense of linear algebra that play the dominant role but other systems, which in general are not bases in the sense of linear algebra. These systems are called *orthonormal bases*. One of their main characteristics is that every vector has a (unique) Fourier expansion in terms of such systems. In this section we summarise the main properties of orthonormal bases of a Hilbert space.

DEFINITION 2.6. Let $S \subset H$. We call S an orthonormal system if every $x \in S$ has norm 1, i.e. $\|x\| = 1$, and any two distinct elements x and y are orthogonal. We call an orthonormal system S an orthonormal basis if it is maximal in the sense that for any orthonormal system T such that $S \subset T$ we have $S = T$.

Using Zorn's lemma one can prove that every Hilbert space and thus every closed subspace of a Hilbert space possesses an orthonormal basis. Moreover, it can be proved that any two orthonormal bases have the same

3. ORTHONORMAL SYSTEMS AND THE FOURIER EXPANSION

cardinality. A Hilbert space having a countable orthonormal basis is called *separable*.

Let us point out that, in the infinite-dimensional case, an orthonormal basis need not be a basis for the Hilbert space in the sense of linear algebra. It is a linearly independent set of elements which, however, need not linearly span the whole space. We will learn shortly, however, that for any orthonormal basis S every vector of the Hilbert space has a Fourier expansion in terms of S which in the finite-dimensional case reduces to a linear combination of elements of S.

We also have, in the infinite-dimensional case, the following analogy with the finite-dimensional case. It is well known that given a finite-dimensional Hilbert space and any linearly independent family $x_1, ... x_n$ there exists an orthonormal set $\{y_1, ... y_n\}$ spanning the same space. Generally, given any countable linearly independent set of vectors T we may construct an orthonormal basis S such that the *closures* of the subspaces spanned by T and S coincide. The reader may find this in the textbooks as the Gram-Schmidt construction.

As already mentioned, one of the chief functions of an orthonormal basis of a Hilbert space H consists in the fact that any $x \in H$ has a Fourier expansion in terms of this orthonormal basis. We need, at this point, to reflect on two things. First, Fourier series are 'infinite sums'. Second, not every Hilbert space is separable, i.e. that orthonormal bases need not be countable. We will thus, if we do not want to restrict ourselves to separable Hilbert spaces, encounter 'infinite sums' with non-denumerable index sets. In the next definitions we explain what we mean by $\sum_{i \in I} x_i$ for any (possibly non-denumerable) index set I.

DEFINITION 2.7. Let H be a normed space with norm $\|.\|$. Let $x_1, x_2, ...$ be a countable sequence of elements of H. Then we say the series $\sum_{n=1}^{\infty} x_n$ converges in the norm with limit x, in symbols $\sum_{n=1}^{\infty} x_n = x$, if for any $\epsilon > 0$ there exists an n_0 such that $\|x - \sum_{n=1}^{m} x_n\| < \epsilon$ for all $m \geq n_0$.

The reader may note that the limit x is unique.

We now say what in this book we mean by the notation $\sum_{i \in I} x_i$.

DEFINITION 2.8. Let I be an arbitrary (index) set I and $x_{i \in I}$ be a family of elements of the normed space H such that the set $\{i \in I \mid x_i \neq 0\}$ is at most countable. Suppose there exists an x such that for any enumeration $x_1, x_2, ...$ of the countable set of non-zero members of $(x_i)_{i \in I}$ we have $\sum_{n=1}^{\infty} x_n = x$. Then we say $\sum_{i \in I} x_i = x$.

PROPOSITION 2.9. *Let $\{e_n, n \in N\}$ be a countable orthonormal system and $x \in H$. Then we have*

$$\sum_{n=1}^{\infty} |\langle x, e_n\rangle|^2 \le \|x\|^2$$

Given any orthonormal system $S = (e_i)_{i \in I}$ and $x \in H$. Then we call the numbers $\langle x, e_i\rangle$ the Fourier coefficients of x (with respect to S).

The above inequality is known as *Bessel's inequality*. Bessel's inequality has the following interesting consequences.

COROLLARY 2.10. *Given any orthonormal system S and $x \in H$. Then $S_x =: \{e \in S \mid \langle x, e\rangle \ne 0\}$ is at most countable.*

This can be seen as follows. Consider the sets $T_{x,n} =: \{e \in S \mid |\langle x, e\rangle| \ge 1/n\}$. By Bessel's inequality these sets are finite and thus $T = \bigcup_{n \in N} S_{x,n}$ is finite or countable.

COROLLARY 2.11. *Given any orthonormal system S and $x \in H$. Then the (countably many non-zero) Fourier coefficients of x form a sequence tending to zero.*

The significance of 2.10 is that for any orthonormal system S we can make sense of a series of the form $\sum_{e \in S} \langle x, e\rangle e$. It's always a 'countable' sum as required in our definition of convergence of a series with a possibly uncountable index set.

THEOREM 2.12. *For any orthonormal system S there exists an orthonormal basis T such that $S \subset T$.*

THEOREM 2.13. *Let S be an orthonormal system. The following conditions are equivalent:*

- *(i) S is an orthonormal basis.*

- *(ii) If $x \in H$ is orthogonal to S, then $x = 0$.*

- *(iii) H is the closure of the subspace spanned by S, i.e. $H = \overline{\text{lin} S}$*

- *(iv) Any $x \in H$ has a unique Fourier expansion in terms of S. This means that $x = \sum_{e \in S} \langle x, e\rangle e$ and this expansion is unique.*

- *(iv) For any $x, y \in H$, $\langle x, y\rangle = \sum_{x \in S} \langle x, e\rangle \langle e, y\rangle$* (**Parseval's Identity**)

Note that Parseval's identity says in particular that $\|x\|^2 = \sum_{e \in S} |\langle x, e\rangle|^2$, if S is an orthonormal basis.

We will also make use of the following well known theorem.

THEOREM 2.14. *Let S be an orthonormal system and suppose that $\sum_{e \in S} |\alpha_e|^2$ is defined. Then there is an $x \in H$ such that $x = \sum_{e \in S} \alpha_e e$ and thus $\alpha_e = \langle x, e\rangle$ for all $e \in S$.*

THEOREM 2.15. *Let S be an orthonormal system, $x \in H$. Then $y :=$ $\sum_{e \in S} \langle x, e \rangle e$ is the projection of x on $\overline{\text{lin} S}$.*

As a consequence we have the following useful observation. Given an orthonormal system $(x_i)_{i \in I}$ and let A be the smallest closed subspace containing it, then the elements of A are precisely those having the form $\sum_{i \in I} \alpha_i x_i$ such that $\sum_{i \in I} |\alpha_i|^2$ is defined.

It can be proved that if two closed subspaces A and B of a Hilbert space are orthogonal, then $A + B$ is closed. Generally, however, this is not true.

We now give an example for two closed subspaces A and B of an infinite-dimensional Hilbert space such that $A + B \neq A \vee B$. For this first note that in an infinite-dimensional Hilbert space we may find two orthonormal sequences $(x_n)_{n \in \mathbb{N}}$ and $(y_m)_{m \in \mathbb{N}}$ such that $x_n \perp y_m$. In fact, this can be seen as follows. Let $(w_n)_{n \in \mathbb{N}}$ be any orthonormal system, then put $x_n = w_{2n}$ and $y_m = w_{2m+1}$. Consider the sequence defined by $z_n := \cos(1/n) x_n + \sin(1/n) y_n$. By the Pythagorean theorem we have $|z_n|^2 = \cos^2(1/n) + \sin^2(1/n) = 1$. Moreover, a straightforward calculation shows that $\langle z_n, z_m \rangle = 0$ for $n \neq m$. Thus the sequence $(z_n)_{n \in \mathbb{N}}$ is an orthonormal system. Let A and B be the smallest closed subspaces containing x_n and z_n respectively. Since $\cos(1/n) \neq 0$ we have $y_n \in A + B$. Now note that $\sum_{n=1}^{\infty} \sin^2(1/n) < \infty$. By 2.14 $y = \sum_{n=1}^{\infty} \sin^2(1/n) y_n$ makes sense. We have $\sin(1/m) = \langle y, y_m \rangle$ and by 2.15 y is an element of $A \vee B$. We now prove that y is not in $A + B$. Suppose $y \in A + B$. Then we have $y = x + z$ with $x \in A$ and $z \in B$. Thus $\sin(1/m) = \langle y, y_m \rangle = \langle x + z, y_m \rangle = \langle z, y_m \rangle = \sum_{n=1}^{\infty} \langle z, z_n \rangle \langle z_n, y_m \rangle = \langle z, z_m \rangle \langle z_m, y_m \rangle = \langle z, z_m \rangle \sin(1/m)$.

The reader should note that the passage from the fifth to the sixth link in the above chain of equations is justified by the continuity of the scalar product and the fact that $\langle z_n, y_m \rangle = \delta_{nm}$ and not by Parseval's identity. It follows that for all m $\langle z, z_m \rangle = 1$, which contradicts the fact that the sequence of Fourier coefficients $\langle z, z_m \rangle$ tends to zero, see 2.11. We have therefore found a y such that $y \in A \vee B$ but $y \notin A + B$.

The above example may also serve as an example of a subspace of an infinite-dimensional Hilbert space which is not closed. It is taken from Halmos' book [27].

4 More lattice theory

DEFINITION 2.16. A lattice is called modular if the modular condition holds:

$$\text{If } A \leq B \text{ then } A \vee (B \wedge C) = (A \vee B) \wedge (A \vee C)$$

If $A \leq B$, then $A \vee B = B$ and thus the modular condition is equivalent to the following:

$$\text{If } A \leq B \text{ then } A \vee (B \wedge C) = B \wedge (A \vee C)$$

DEFINITION 2.17. *An orthocomplemented lattice is called an orthomodular lattice if the orthomodular condition holds: If $A \leq B$ and $A^\perp \leq C$ then $A \vee (B \wedge C) = (A \vee B) \wedge (A \vee C)$ In fact, in this case we have $A \vee (B \wedge C) = B$*

There are various equivalent definitions of orthomodularity, see for instance Redei's book [52].

The following version will be of use later in the book.

PROPOSITION 2.18. *An orthocomplemented lattice is orthomodular iff $A \leq B$ implies $B = A \vee (A^\perp \wedge B)$.*

Note that distributivity implies modularity and modularity implies orthomodularity. The converse implications do not hold.

We write $A < B$ for $A \leq B$ and $A \neq B$.

DEFINITION 2.19. *Given a lattice L and $A, B \in L$. We say that B covers A if $A < B$ and $A < C < B$ is satisfied by no C.*

An element $A \in L$ is called an atom if it covers 0. L is called atomic if for any $B \in L$ there exists an atom A such that $A \leq B$. L is called atomistic if any element B is equal to the least upper bound of those atoms A satisfying $A \leq B$. We say that L has the covering property if the following holds: if P is an an atom and $A \in L$, then $A \vee P$ covers A. We say that A commutes with B or synonymously A is compatible with B if $A = (A \wedge B) \vee (A \wedge B^\perp)$. Call the set of all elements of L commuting with all others the centre of L. If the centre of L consists of 0 and 1 only we call L irreducible.

Remark: In the lattice of projections of a Hilbert space two projections commute in the sense above iff they commute as operators.

DEFINITION 2.20. *Given any poset $\mathcal{P}$ we may define the notion of a chain and that of the length of a chain in a straightforward way. We then define the height of $\mathcal{P}$ denoted by $h(P)$ as the supremum over the lengths of all chains of $\mathcal{P}$ minus 1.*

PROPOSITION 2.21. *In a complete orthocomplemented lattice we have the de Morgan rules:*

- $(\bigvee_i A_i)^\perp = \bigwedge_i A_i^\perp$
- $(\bigwedge_i A_i)^\top = \bigvee A_i^\perp$

where the symbols $\bigwedge$ and $\bigvee$ denote the greatest lower bound and the least upper bound respectively.

4. MORE LATTICE THEORY

In the sequel we will use the term polynomial. For this note that $\wedge$ and $\vee$, $\perp$ may be viewed as algebraic operations. The term 'polynomial' may then be defined as usual in (universal) algebra. By a (lattice) conditional we mean a polynomial in two variables. For instance, given two elements A and B of an orthocomplemented lattice, then the polynomial $S(A, B) =: A^\perp \vee B$ represents a conditional, namely 'material implication'.

LEMMA 2.22. *Let L be an orthocomplemented lattice. Suppose there exists a conditional $S(A, B)$ satisfying the following condition.*

$$A \wedge C \leq B \text{ iff } C \leq S(A, B$$

Then L is a Boolean algebra.

Proof. For the sake of convenience we write $A \rightsquigarrow B$ for $S(A, B)$. Given any elements A, B, C. We need to show that $((A \vee B) \wedge C) = (A \wedge C) \vee (B \wedge C)$. For this it suffices to show that $((A \wedge B) \vee C) \leq (A \wedge C) \vee (B \wedge C)$. We have

$$A \wedge C) \leq (A \wedge C) \vee (B \wedge C)$$

and

$$(B \wedge C) \leq (A \wedge C) \vee (B \wedge C)$$

Using the condition on $S(A, B)$ we obtain

$$A \leq (C \rightsquigarrow ((A \wedge C) \vee (B \wedge C)))$$

and

$$B \leq (C \rightsquigarrow ((A \wedge C) \vee (B \wedge C)))$$

Hence

$$(A \vee B) \leq ((C \rightsquigarrow ((A \wedge C) \vee (B \wedge C)))$$

Applying the condition on $S(A, B)$ in the other direction we get

$$(A \vee B) \wedge C \leq (A \vee B) \wedge (A \vee B)$$

This is what we wanted to prove. ∎

THEOREM 2.23 (Mittelstaedt). *Let L be an orthocomplemented lattice with orthocomplementation $\perp$. Then L is orthomodular if there exists a conditional $S(A, B)$ such that the following conditions are satisfied.*

$$(i)\ A \wedge S(A,B) \leq B$$

$$(ii)\ A \wedge C \leq B \text{ implies } A^\perp \vee (A \wedge C) \leq S(A,B)$$

A conditional satisfying the above conditions is unique, namely

$$S(A,B) = A^\perp \vee (A \wedge B).$$

L is a Boolean algebra if the above conditions are satisfied by 'material implication', i.e. $S(A,B) = A^\perp \vee B$.

We denote $A^\perp \vee (A \wedge B)$ by $A \leadsto_s B$.

Proof. Assume there exists a conditional satisfying (i) and (ii). We need to show that L is orthomodular. So let $B \leq A$ and $C \leq A^\perp$. Then we have

$$B = B \wedge A \leq A^\perp \vee (A \wedge B) = A \leadsto_s B)$$

Moreover we have

$$C \leq A^\perp \leq A^\perp \vee (A \wedge B) = A \leadsto_s B$$

Hence

$$B \vee C \leq A \leadsto_s B$$

and thus

$$A \wedge (B \vee C) \leq A \wedge (A \leadsto_s B)$$

By condition (i) we have

$$A \wedge (A \leadsto_s B) \leq B$$

It follows that

$$A \wedge (B \vee C) \leq B$$

and by the definition of orthomodularity this means that L is orthomodular.

For the proof of uniqueness let $S'(A,B)$ be any conditional satisfying (i) and (ii). We can then assume orthomodularity of $\mathcal{L}$. Putting $C = B$ we get by (ii) that $A \leadsto_s B \leq S'(A,B)$. Since $S'(A,B)$ satisfies (i) it follows that $A \wedge S'(A,B) \leq (A \wedge B) \leq A^\perp \vee (A \wedge B) = A \leadsto_s B$. Hence $A^\perp \vee (A \wedge S'(A,B)) \leq A \leadsto_s B)$. Considering that $A^\perp \leq S'(A,B)$ we have by orthomodularity that $A^\perp \vee (A \wedge S'(A,B)) = S'(A,B)$. It follows that $S'(A,B) \leq A \leadsto_s B$. $S'(A,B)$ and $A \leadsto_s B$ are thus equal.

We still need to prove that if conditions (i) and (ii) are satisfied by $S(A,B) =: A^\perp \vee B$, L is a Boolean algebra. For this we have to verify the condition in the preceding lemma. First note that in this case $A^\perp \vee (A \wedge B) = A^\perp \vee B$. So assume $A \wedge B \leq C$. Then it follows from condition (ii) that $A^\perp \vee C \leq A^\perp \vee B$. Hence $C \leq A^\perp \vee B$. For the other direction assume $C \leq A^\perp \vee B$. Then we have $(A \wedge C) \leq A \wedge (A^\perp \vee B)$. By condition (i) we have $A \wedge (A^\perp \vee B) \leq B$. Thus $(A \wedge C) \leq B$. This completes the proof. ∎

5 The lattice of closed subspaces and projections of an orthomodular space

In this section we do not restrict ourselves to Hilbert spaces. Rather we always have in mind the more general case of an *orthomodular space*. The concept of an orthomodular space is more general than that of a Hilbert space, but it suffices for many purposes. Unless explicitly mentioned otherwise the spaces under consideration in this section are orthomodular spaces.

DEFINITION 2.24. Let K be a (not necessarily commutative) field with an involution τ, i.e. a function $\tau : K \to K$ such that

$$\tau(a+b) = \tau(a) + \tau(b), \ \tau(ab) = \tau(b)\tau(a), \ \tau\tau(a) = a$$

Let H be a vector space over K and $\langle . \rangle : H \times H \to K$ be a Hermitian form on H, i.e. $\langle . \rangle$ satisfies

$$\langle ax + by, z \rangle = a\langle x, z \rangle + b\langle y, z \rangle$$

$$\langle z, ax + by \rangle = \langle z, x \rangle \tau(a) + \langle z, y \rangle \tau(b)$$

$$\langle x, z \rangle = \tau(\langle z, x \rangle)$$

$$\langle x, x \rangle = 0 \text{ implies } x = 0,$$

then call the pair $\langle H, \langle . \rangle \rangle$ a *Hermitian space*.

Define the concepts of *orthogonality* of vectors and the orthogonal complement $A^\perp$ of a subspace A as in the case of Hilbert spaces. Call a subspace A *closed* iff $A = A^{\perp\perp}$.

DEFINITION 2.25. Call a Hermitian space $\langle H, \langle . \rangle \rangle$ an *orthomodular space* iff for every closed subspace A we have

$$H = A \oplus A^\perp$$

Every Hilbert space H is an orthomodular space. A subspace of H is closed in the sense of the topology of a Hilbert space iff it is closed in the sense of an orthomodular space as defined above. Moreover, let us remark that given an orthomodular space H and a closed subspace A of H , then A is itself an orthomodular space with the Hermitian form of H restricted to A. This corresponds to the fact that every closed subspace of a Hilbert space is itself a Hilbert space.

Again, we denote the set of closed subspaces of H by $Sub(H)$. It is obvious that set inclusion is a partial order on $Sub(H)$. Moreover, the intersection of closed subspaces is a closed subspace. Clearly, H itself is closed. Given $A, B \in Sub(H)$, then $A \cap B$ is the greatest lower bound of A and B and the lowest upper bound of A and B denoted by $A \vee B$ is given by the smallest closed subspace containing A and B. Orthogonal complement formation is an orthocomplementation of $Sub(H)$. The zero space is the null element of the lattice and H is the unit element. $Sub(H)$ is thus an orthocomplemented lattice.

Since for every closed subspace A and any $x \in H$ we have by definition a unique decomposition of x of the form $x = y + z$ with $y \in A$ and $z \in A^\perp$ we may, as in the case of a Hilbert space, associate a projection to every closed subspace A denoted by P_A or, context permitting, just A namely by putting $P_A x = y$ where y is as in the above decomposition. We observe, as in the case of a Hilbert space, that $A = \{x \in H \mid P_A x = x\}$. Again we have a one-to one correspondence between closed subspaces and projections since $P_A = P_B$ iff $A = B$. Denote the set of projections by $PRO(H)$.

We have the following connection between closed subspaces and the corresponding projections. $A \subset B$ iff $P_B P_A = P_A$. This is seen as follows. For the direction from right to left let $x \in A$. Then we have $P_A x = x$ and thus, since $P_B P_A(x) = P_A x$, $P_B x = x$. Hence $x \in B$. For the other direction assume $A \subset B$. Let $x \in H$. We have $P_A x \in A$ and thus by the hypothesis $P_A x \in B$. Hence $P_B(P_A x) = P_A x$. Define a binary relation $\leq$ in $PRO(H)$ by $P_A \leq P_B$ iff $P_B P_A = P_A$. Moreover observe that the mapping $*: PRO(H) \to PRO(H)$ given by $P_A{}^* = P_{A^*}$ is well defined. Then $\langle PRO(H), \leq, * \rangle$ is an orthocomplemented lattice (ortho-) isomorphic to the lattice of closed subspaces via the mapping $A \mapsto P_A$.

In the sequel we mean ortho-isomorphism (automorphism) whenever we use the term isomorphism (automorphism).

The reader may, in view of the fact that the lattice of closed subspaces of an orthomodular space and thus of a Hilbert space and the lattice of projections are isomorphic, think that these two two isomorphic structures are equally suited in every respect, in particular for logical purposes. That this is not the case is even essential for our purpose. Namely, subspaces and projections of an orthomodular space are conceptually distinct entities. The fact that projections are operators allows for a dynamic view of propositions if represented by projections.

The following proposition generalises an observation made by Hardegree in [28] in connection with Hilbert spaces.

PROPOSITION 2.26. *Let H be an orthomodular space, $x \in H$, $A, B \in Sub(H)$. Then $Ax \in B$ iff $x \in A^\perp \vee (A \wedge B)$.*

Proof. First note that the closed subspaces $A^\perp$ and $A \wedge B$ are orthogonal. Then we have $A^\perp \vee (A \wedge B) = A^\perp \oplus (A \wedge B)$.

For the direction from left to right let $x = y + z$ be the unique decomposition of x with respect to A and $A^\perp$, i.e. $y \in A$ and $z \in A^\perp$. We have $Ax = y$. The hypothesis says that $y \in B$. Thus $y \in A \wedge B$. It follows that $x \in A^\perp \oplus (A \wedge B)$.

For the direction from right to left observe $A^\perp \oplus (A \wedge B)$ is again an orthomodular space with the Hermitian form properly restricted. We have thus, in addition to the above decomposition, a decomposition $x = y_1 + z_1$ with $y_1 \in A$ and $z_1 \in A \wedge B$. Since the decomposition is unique we have $y = y_1$ and $z = z_1$. It follows that $Ax = y = y_1 \in B$. ∎

We call, for historical reasons, a lattice a Hilbert lattice if it is isomorphic to the lattice of closed subspaces of an orthomodular space.

THEOREM 2.27. *A Hilbert lattice is an atomistic, complete, orthomodular irreducible lattice having the covering property.*

The following theorem is *Piron's Representation Theorem*, see [48].

THEOREM 2.28 (Piron). *An ortholattice $\mathcal{L}$ of height ≥ 4 is a Hilbert lattice iff it is atomistic, complete, irreducible, orthomodular and satisfies the covering property.*

For the proof of the following interesting theorem see [2]

THEOREM 2.29 (Amemiya-Araki). *A Hilbert space H is finite dimensional iff $Sub(H)$ is modular.*

It is an important fact that for any infinite-dimensional Hilbert space H, $Sub(H)$ is orthomodular but *not* modular.

6 Characterising classical Hilbert lattices

In quantum mechanics we are (primarily) concerned with infinite-dimensional Hilbert spaces. We define, for historical reasons, a *classical Hilbert lattice* to be a lattice isomorphic to the lattice of closed subspaces of an infinite-dimensional Hilbert space over the real numbers, the complex numbers or the quaternions. Recall that we have already defined a *Hilbert lattice* to be a lattice isomorphic to the lattice of closed subspaces of some orthomodular space. For these lattices we have Piron's representation theorem 2.28, which characterises Hilbert lattices of height at least 4 among ortholattices.

In this section we give a characterisation of classical Hilbert lattices among ortholattices which was first presented by Engesser in [16]. For this purpose we need, apart from Piron's theorem, two deep theorems of modern Hilbert space theory, namely the theorems of Solèr 2.30 and Mayet 2.33, which we state below. Mayet's theorem heavily relies on a theorem of Wigner, which we state too. In his pioneering paper [36] Keller settled a long standing question, namely the question whether every infinite-dimensional orthomodular space is already a Hilbert space. Keller's construction of a counter example settled the question in the negative. This, however, posed another problem, namely the problem of characterising those orthomodular spaces that are in fact Hilbert spaces. This problem was solved by Maria Pia Solèr in [59].

THEOREM 2.30 (Solèr). *Let $\langle H, \langle, \rangle \rangle$ be an orthomodular space over K and let $c \in K$. Suppose there exists an infinite family $(x_i)_{i \in I}$ of pairwise orthogonal elements of H such that for all $i \in I$, $\langle x_i, x_i \rangle = c$. Then K must be the (skew-) field of the real numbers, the complex numbers or the quaternions and H is an infinite-dimensional Hilbert space.*

DEFINITION 2.31. Let H_1 and H_2 be two orthomodular spaces and $\sigma : H_1 \to H_2$ be a bijective map. We say that σ is a semi-unitary map iff the following conditions are satisfied.

- For any $x, y \in H_1$, $\sigma(x+y) = \sigma(x) + \sigma(y)$.

- There exists an automorphism ρ of K such that, for any $\lambda \in K$ and any $x \in H_1$, we have $\sigma(\lambda x) = \rho(\lambda)(\sigma x)$.

- There exists $\lambda_\sigma \in K$ such that, for any $x, y \in H_1$, we have $\langle \sigma(x), \sigma(y) \rangle = \rho(\langle x, y \rangle)\lambda_\sigma$.

If, moreover, we have $\rho = id_K$ and $\lambda_\sigma = 1$, we say that σ is unitary.

THEOREM 2.32 (Wigner). *Let H_1 and H_2 be orthomodular spaces of dimension at least 3. Then every ortholattice isomorphism $f : Sub(H1) \to Sub(H_2)$ is induced by some semi-unitary map.*

We need the following result by Mayet which, essentially, is a consequence of Wigner's theorem.

THEOREM 2.33 (Mayet). *Let H be an orthomodular space of dimension at least 3 and let $X \in Sub(H)$ of dimension at least 2. Let f be an automorphism of $Sub(H)$ whose restriction to $[0, X]$ is the identical map. Then there*

6. CHARACTERISING CLASSICAL HILBERT LATTICES

exists a unique unitary operator σ on H inducing f such that the restriction of σ to X is the identical map.

Solèr's theorem characterises Hilbert spaces among orthomodular spaces. We are interested in a characterisation of classical Hilbert lattices among ortholattices.

The characterisation we give is in terms of a symmetry property (see [16].

For a given ortholattice L we call two atoms σ_1 and σ_2 orthogonal if $\sigma_1 \leq \sigma_2^\perp$. This relation is readily seen to be symmetric.

DEFINITION 2.34.
Let L be a complete ortholattice and let $\Delta = (\sigma_i)_{i \in I}$ be an infinite pairwise orthogonal family of atoms of L. We say that L satisfies the symmetry property (synonymously: is symmetric) with respect to Δ iff the following holds. For any permutation $f : I \to I$ there exists an ortholattice automorphism ρ_f of L with the following properties.

- ρ_f extends f, i.e. $\rho_f(\sigma_i) = \sigma_{f(i)}$) for any $i \in I$.

- If the set J of those elements of I which are left fixed by f is non-empty, ρ_f induces the identical map on $[0, A]$, where A denotes the least upper bound of the family $(\sigma_j)_{j \in J}$.

We say that L is symmetric iff there exists an infinite pairwise orthogonal family Δ of atoms of L such that L is symmetric with respect to Δ.

We have the following characterisation theorem.

THEOREM 2.35 (Engesser). *A Hilbert lattice is a classical Hilbert lattice iff it is symmetric.*

Proof. Let us first verify that for a given infinite-dimensional classical Hilbert space H, $Sub(H)$ is symmetric. To see this consider an orthonormal basis $(x_i)_{i \in I}$ of H. Then the family of one-dimensional subspaces $(\langle x_i \rangle)_{i \in I}$ is an infinite orthogonal system of atoms of $Sub(H)$. Let $f : I \to I$ be any permutation of I. Recall that $x = \sum_{i \in I} \langle x, x_i \rangle x_i$ by 2.13. Define the map φ_f as follows. For $x = \sum_{i \in I} \langle x, x_i \rangle x_i$ put $\varphi_f(x) := \sum_{i \in I} \langle x, x_{f^{-1}(i)} \rangle x_i$. φ is well defined in view of theorems by 2.13 and 2.14. For any $i \in I$ we have $\varphi_f(x_i) = x_{f(i)}$. Moreover, φ_f is unitary, since for any $x, y \in H$ we have by Parseval's identity $\langle \varphi_f(x), \varphi_f(y) \rangle = \sum_{i \in I} \langle x, x_{f^{-1}(i)} \rangle \overline{\langle y, x_{f^{-1}(i)} \rangle} = \sum_{i \in I} \langle x, x_i \rangle \overline{\langle y, x_i \rangle} = \langle x, y \rangle$. Suppose $\{i \mid f(i) = i\}$ is non-empty and denote by X the smallest closed subspace containing $\{x_i \mid f(i) = i\}$. X is the smallest closed subspace containing $\{\langle x_i \rangle \mid f(i) = i\}$ and φ_f induces the identity on X. For the latter claim observe that φ_f induces the identity on the subspace spanned by $\{x_i \mid f(i) = i\}$ and and X is the closure of that subspace. Since φ_f is continuous, it induces the identity on X too. φ_f thus induces an ortholattice automorphism ρ_f on $Sub(H)$ such that for any $i \in I$, $\rho_f(\langle x_i \rangle) = \langle x_{f(i)} \rangle$. Clearly, ρ_f induces the identical map on $[0, X]$. Thus symmetry of $Sub(H)$ is proved.

For the other direction note that the symmetry property implies infinite height. By Piron's Representation Theorem it therefore suffices to show that any orthomodular space H such that $Sub(H)$ has the symmetry property is an infinite-dimensional classical Hilbert space. So let $(\langle x_i \rangle)_{i \in I}$ be an infinite orthogonal family with respect to which $Sub(H)$ is symmetric. Let $i_0 \in I$. For any $j \in I, i_0 \neq j$ consider the permutation f_j of I defined as follows.

$$f_j(i_0) = j, \ f_j(j) = i_0, \ f_j(i) = i \text{ else.}$$

Denote by X the smallest closed closed subspace of X containing $\langle x_i \rangle$ for all $i \in I$. X is infinite-dimensional. By symmetry there exists an automorphism ρ_j of $Sub(H)$ inducing the identity on $[0, X]$ such that for all $i \in I$, $\rho_j(\langle x_i \rangle) = \langle x_{f_j(i)} \rangle$. So, by Mayet's theorem, ρ_j is induced by some unitary map φ_j. Put $y_j =: \varphi_j(x_{i_0})$ for $j \neq i_0$ and $y_{i_0} = x_{i_0}$. Then, since φ_j is unitary, the family $(y_j)_{j \in I}$ is a family as required in Solèr's theorem. It follows by Solèr's theorem that H must be an infinite-dimensional classical Hilbert space. ∎

As a corollary we get the following theorem, which gives another characterisation of Hilbert spaces among orthomodular spaces.

THEOREM 2.36. *Let $\langle H, \langle . \rangle \rangle$ be an orthomodular space over K. Then the following conditions are equivalent.*

- *There exists an infinite family $(x_i)_{i \in I}$ of pairwise orthogonal elements of H and a non-zero $c \in H$ such that for all $i \in I$ we have $\langle x_i, x_i \rangle = c$.*

- *$Sub(H)$ is symmetric.*

- *H is an infinite-dimensional classical Hilbert space.*

CHAPTER 3

BASICS OF THE FORMALISM OF QUANTUM MECHANICS

1 Some history

It is a common experience among students of physics that their first course on quantum mechanics comes as a sort of shock. In such a course the student is confronted with a discipline that does not display the pattern he is used to from the physical theories he has already mastered such as Newtonian mechanics, electrodynamics, special relativity. The student must come to terms with the fact that, in contrast to these classical physical theories, quantum mechanics is essentially a mathematical formalism. He is taught how to make use of this formalism in order to calculate certain physical quantities such as the energy levels of the electron in the hydrogen atom. The success of the formalism of quantum mechanics is unique in the history of science yielding the correct results with unprecedented precision for a vast range of phenomena which were entirely untractable in classical physics. This is the reason for the wide spread slogan that quantum mechanics is the "most successful physical theory ever".

However, the student's question "Why this formalism? Where does it come from?" gets normally, if at all, an evasive and unsatisfactory answer. The plain truth is that this formalism is the result of guesswork, ingenious guesswork admittedly.

The first version of the formalism of quantum mechanics has become known as *matrix mechanics*. The first and already crucial step in this process of guessing was taken in June 1925 by Werner Heisenberg, a then 23 year old post-doc, in his famous paper [29]. Essentially, the discovery was that physical quantities such as energy, momentum etc. are to be represented by infinite matrices in such a way that the possible values a physical quantity can assume are given by the eigenvalues of the corresponding matrix. The matrices representing physical quantities were not required to commute and non-commutation of matrices was to be regarded as "non-simultaneous measurability" of the corresponding physical quantities.

It is interesting to note that Heisenberg did not explicitly use the concept of a matrix. In fact, he was not even familiar with that concept at the time. The concept of a matrix was then not part of a mathematician's standard background let alone a physicist's. The technical tools used in the paper were somehow vague as a sort of 'schemata of numbers' which can be combined (multiplied) in a certain way. It was Heisenberg's teacher Max Born who, after Heisenberg's paper had appeared, realised that these mathematical schemata were actually (infinite) matrices and Heisenberg's way of combining them was actually matrix multiplication. In the joint paper [8] by Born and Jordan, which appeared only a few months after Heisenberg's paper, the first precise account of matrix mechanics was given for systems with one degree of freedom. In particular it contains the first mathematically precise statement of Heisenberg's Uncertainty Principle: If P is the matrix representing the position of a particle and Q is the matrix representing its momentum, we have

$$PQ - QP = \frac{h}{\pi i}$$

In the subsequent classic paper by Born-Heisenberg-Jordan [9] the formalism of matrix mechanics was fully established. The concept of a Hilbert space, however, still had no place in this. This had to wait for John von Neumann's work in the late twenties (see [60], [61]) and in particular his classic book [62] of 1932, in which the formalism received its final elegant shape.

It is interesting that similar guesswork led to Schrödinger's *wave mechanics*, which later on was proved to be equivalent to matrix mechanics.

In 1926 Erwin Schrödinger published his famous paper [56], which, essentially earned him the Nobel Prize as did Heisenberg's paper [29] for Heisenberg. In that paper he claimed and proved to have found a (partial) differential equation with a remarkable property. This differential equation has become known as the Schrödinger equation, which nowadays is probably the most famous equation of physics. And what was so remarkable about it? Well, the claim Schrödinger made and proved was that his equation had square integrable solutions exactly for those eigenvalues that correspond to the experimentally found energy levels which the electron in the hydrogen atom can assume. This is all Schrödinger claimed and proved. The solution of Schrödinger's equation usually denoted by $\Psi(x, y, z)$ is, in the stationary case, a complex valued functions of the three spatial coordinates x, y, z and in the non-stationary case also of time t. Schrödinger had, at the time when his paper appeared, no idea of the physical meaning of the function Ψ. The nowadays generally accepted physical interpretation given to the Ψ function by Born in [7] is that $\mid \psi \mid^2 dxdydz$ represents the probability with which the particle can be found in the infinitesimal volume $dxdydz$. So, according to this interpretation, the solutions of the Schrödinger equation are *probability*

2. HERMITIAN OPERATORS

waves. It is an irony of the history of science that Schrödinger himself could never make friends with that interpretation of the function Ψ.

To summarise, the core of quantum mechanics is a formalism, and this formalism was *guessed*. It always works with amazing precision. One of the aims we pursue in this book is to put logic to good use in order to shed light on this formalism.

2 Hermitian operators

We will see that certain operators on Hilbert spaces play a vital role in the formalism of quantum mechanics. These operators are called *self-adjoint* and, if they are bounded, *Hermitian operators*. Given a Hilbert space H. Then we call any linear map of H into itself an operator of H. Given an operator T on H and let T^* be such that for all $x, y \in H$ we have $\langle Tx, y \rangle = \langle x, T^*y \rangle$. Then we call T^*, which is unique, the adjoint of T. We call T self-adjoint if $T = T^*$. We call an operator T bounded if there exists a positive real number c such that for all $x \in H$ we have $\|Tx\| \leq c\|x\|$. One can then prove that an operator is bounded iff it is continuous. It can also be proved that any bounded operator T has a unique adjoint, which is also bounded.

DEFINITION 3.1. Call an operator T Hermitian if is bounded and self-adjoint. Call a bounded operator T unitary if T is bijective and for any x and y we have $\langle Tx, Ty \rangle = \langle x, y \rangle$

It can then be proved that a unitary operator may also be defined as a bijective Hermitian operator the adjoint of which is its inverse.

PROPOSITION 3.2.

Let T be an Hermitian operator. Then

- *There exists an orthonormal basis of eigenvectors of T.*

- *The eigenvalues of T are real numbers.*

From this it follows by 2.13 that, given a Hermitian operator T, any vector has a Fourier expansion in terms of eigenvectors of T.

PROPOSITION 3.3. *Let S and T be Hermitian operators. Then the following conditions are equivalent*

- *S and T commute, i.e. $ST = TS$*

- *ST is a Hermitian operator.*

- *There exists an orthonormal basis which is a family of eigenvectors common to both S and T.*

3 Postulates of quantum mechanics

In this section we describe the basics of the formalism of quantum mechanics without giving a complete description of the Hilbert space formalism and its subtleties. We restrict ourselves to those postulates which the reader must know in order to understand the subsequent chapters. For a complete description the reader is referred to [11].

The mathematical formalism of quantum mechanics provides the answers to the following questions.

- 1) How is the state of a physical system $\mathcal{S}$ represented mathematically?

- 2) How are physical quantities represented mathematically?

- 3) What are the possible outcomes of a measurement of a given physical quantity?

- 4) How does a given state change in the process of measurement?

- 5) How does the system evolve over time?

- 6) How is the mathematical representation of a composite system built from the mathematical representations of the component systems?

POSTULATE 3.4. With a given a physical system $\mathcal{S}$ there is associated a Hilbert space H over the complex numbers such that at any fixed time there exists a one-to-one correspondence between the states of $\mathcal{S}$ and the rays of H.

POSTULATE 3.5. Every physical quantity $\mathcal{A}$ of $\mathcal{S}$ corresponds to a self-adjoint operator A of H.[1] We call the operator A the observable corresponding to $\mathcal{A}$.

We call the operator A the observable corresponding to $\mathcal{A}$.

POSTULATE 3.6. The possible results of a measurement of a physical quantity $\mathcal{A}$ are eigenvalues of the corresponding observable A.

Recall that by 3.2 the eigenvalues of A are real numbers.

Note that Hermitian operators may have a continuous set of eigenvalues rather than a discrete one. For simplicity we formulate the following postulates only in the discrete case. In the discrete case we again have to distinguish between two cases, the non-degenerate case and the degenerate

---

[1] We do not use the term Hermitian operator here, since we defined a Hermitian operator to be bounded as usual in Functional Analysis. There are observables in quantum mechanics, however, which are not bounded such as those of position and momentum.

3. POSTULATES OF QUANTUM MECHANICS

case. We say that we have the non-degenerate case if all eigenspaces are one-dimensional, else we say t at we have the degenerate case.

For simplicity we formulate the following postulate for the non-degenerate case only.

POSTULATE 3.7 (non-degenerate case). Suppose the physical quantity $\mathcal{A}$ is measured in state $\langle x \rangle$. Then the probability $P(a_n, \langle x \rangle)$ of obtaining the (non-degenerate) eigenvalue a_n is calculated as follows. Take any normalised vector x_0 of $\langle x \rangle$

$$P(a_n, \langle x \rangle) = |\langle y_n, x_0 \rangle|^2,$$

where y_n is any normalised eigenvector corresponding to the eigenvalue a_n.

Note that the above postulate is well stated because any two normalised vectors of a ray differ only by a complex factor of modulus one.

The following postulate is called the *Projection Postulate*

POSTULATE 3.8. Suppose the system is in state $\langle x \rangle$ and a measurement of the quantity $\mathcal{A}$ is performed in this state. Suppose we get as a result of this measurement the value a. By Postulate 2, a is an eigenvalue of A. Denote by P_a the projection onto the eigenspace corresponding to the eigenvalue a. Then the state of the system 'after measurement' is $\langle y \rangle$ where $y = P_a(x)$. Note that this does not depend on the choice of x.

Assume we have two commuting (Hermitian) observables A and B: $AB = BA$. By 3.3 we then know that A and B have a common complete orthogonal system of eigenvectors. We may then, in view of the projection postulate, say that the physical quantities represented by A and B are 'simultaneously measurable'. If A and B do not commute, they do not possess a common system of eigenvectors and the corresponding physical quantities are thus not 'simultaneously measurable'. In this case we also say that we have an *uncertainty relation* between these physical quantities.

We will not be concerned with the temporal evolution of systems so that we need not go into that part of the formalism in depth. Just note that the temporal evolution of a state is governed by the *Schrödinger equation*.

$$\tfrac{h}{\pi i} \tfrac{d}{dt} \Psi(t) = H \Psi(t)$$

where H is the operator corresponding to the energy of the system.[2]

---

[2] Note that differentiation in the above context make sense. Recall that a Hilbert space is a Banach space and we can define differentiation of functions of a real variable with values in a Banach space in a way analogous to the way we define differentiation of real valued functions.

4 Combining systems

In order to formulate the postulate concerning the combination of systems we need the concept of a tensor product of Hilbert spaces. A mathematically rigorous introduction of this concept would be beyond the scope of this book. We just need a few properties of the tensor product which we will state below.

Given two Hilbert spaces H_1 and H_2. There exists a Hilbert space denoted by $H_1 \otimes H_2$ called the tensor product of H_1 and H_2 with the following properties. There exists a mapping $g : H_1 \times H_2 :\to H_1 \otimes H_2$

$$(v, w) \mapsto v \otimes w$$

with the following properties:

(1) g is bilinear. This means that

$$(v_1 + v_2) \otimes w = v_1 \otimes v_2 + v_2 \otimes w$$

$$v \otimes (w_1 + w_2) = v \otimes w_1 + v \otimes w_2$$

$$(\lambda v) \otimes w = \lambda(v \otimes w) = v \otimes (\lambda w)$$

From this it follows that

$$v \otimes 0 = 0 = 0 \otimes v$$

(2) Any element of $H_1 \otimes H_2$ is a linear combination of elements of the form $v \otimes w$.

(3) We have the following connection between the scalar product in the tensor product and those in the component Hilbert spaces:

$$\langle v \otimes v', w \otimes w' \rangle = \langle v, v' \rangle \langle w, w' \rangle$$

POSTULATE 3.9. Given two physical systems $\mathcal{S}_1$ and $\mathcal{S}_2$ with the Hilbert spaces H_1 and H_2 respectively. Then the Hilbert space corresponding to the combined system is given by $H_1 \otimes H_2$.

CHAPTER 4

BIRKHOFF-VON NEUMANN 1936

It was in 1932 that John von Neumann's classic book *Mathematische Grundlagen der Quantenmechanik* [62] appeared in print. In that book the mathematical formalism of quantum mechanics received its elegant modern form with Hilbert space as its core mathematical structure. In 1936 John von Neumann published a paper entitled " The logic of quantum mechanics" [4] jointly with the Harvard mathematician Garret Birkhoff. This paper marks the birth of what has become known as quantum logic.

Birkhoff and von Neumann's seminal work has something in common with Keynes' famous book "General Theory of Employment, Interest and Money", which initiated Keynesianism. Both works are widely quoted, but not all of those quoting these works have actually studied them in depth. Possibly this is due to the fact that these works were not only highly influential in their respective fields but that they are also not easily readable, to say the least.

The paper is frequently quoted as introducing the lattice of closed subspaces of a Hilbert space as the core algebraic structure of the 'logic of quantum mechanics'. This is a poor description of this important work. In this section we attempt a detailed analysis of this paper. This is, in view of the importance of this work, an end in itself, but it is also fruitful for making the connection with the approach adopted in this book. In this section we will analyse and in a way reconstruct and interpret this work quoting extensively from the paper itself. We also occasionally use Birkhoff and von Neumann's notation.

1 Structure of the paper

The paper starts with an Introduction and ends with an Appendix. Its core is divided into three parts:
  (1) Physical Background
  (2) Algebraic Analysis
  (3) Conclusions

We try to make transparent the train of thought of the paper and the peculiar reasoning, which, as pointed out by the authors themselves, is not free from heuristic features. Our special interest concerns "Physical Background", where the connection between the 'logic of quantum mechanics' and closed subspaces of a Hilbert space is made.

2 Novel logical notions in quantum mechanics.

The paper starts, in the Introduction, as follows: "One of the aspects of quantum theory which has attracted the most general attention is the novelty of the logical notions it presupposes. It asserts that even a complete mathematical description of a physical system S does not in general enable one to to predict with certainty the result of an experiment on S, and that in particular one can never predict with certainty both the position and the momentum of S (Heisenberg's Uncertainty Principle). It further asserts that most pairs of observations cannot be made on S simultaneously (Principle of Non-commutativity of Observations)."

This is worth reflecting on. The authors start by saying that quantum theory presupposes new logical notions unfamiliar from classical logic. Then two examples for such 'novel logical notions' are given. The first example is Heisenberg's Uncertainty Principle, and the second example is what the authors call the Principle of Non-Commutativity of Observations. At first glance these principles seem to be purely physical in nature. What is interesting here is that Birkhoff and von Neumann obviously consider these principles not just as novel physical principles but also as novel *logical* notions. This is in fact remarkable.

The paper continues as follows: "The object of the present paper is to discover what logical structure one may hope to find in physical theories which, like quantum mechanics, do not conform to classical logic. Our main conclusion, based on admittedly heuristic arguments, is that one can reasonably expect to find a calculus of propositions which is formally indistinguishable from the calculus of linear subspaces with respect to set products, linear sums and orthogonal complements -and resembles the usual calculus of propositions with respect to *and*, *or*, and *not*".

Again, this passage is worth reflecting on. What do Birkhoff and von Neumann mean by saying that quantum mechanics does not conform to classical logic? Why does it not conform to classical logic? Is it not true that physicists use classical logic in reasoning about quantum systems? In fact, as already mentioned, Popper for instance does not share the view that quantum mechanics does not 'conform' to classical logic. In [49] he says: "...physical theories, including quantum mechanics, do conform to classical logic, even according to Birkhoff and von Neumann's proposal."

There is a letter written by von Neumann to Birkhoff (see [53]) dated November 2 1935 — the paper was received by Annals of Mathematics on

2. NOVEL LOGICAL NOTIONS IN QUANTUM MECHANICS.

April 4, 1936 — which may cast light on this. John von Neumann writes: "Looking at the paper now I see, that I forget to say this...: That while common logics did apply to quantum mechanics, if the notion of simultaneous measurability is introduced as an auxiliary notion, we wished to construct a logical system, which applies directly to quantum mechanics — without any extraneous secondary notions like simultaneous measurability. And in order to have such a consequent, one-piece system of logics, we must change the classical class calculus of logics."

This passage is crucial to the understanding of the Birkhoff-von Neumann enterprise. From the modern point of view one can reconstruct this as follows. Given a (formal) language that permits us to make statements about classical mechanics. Then these statements (propositions) are expected to obey classical logic. In such a language there is no need for talking about compatibility or incompatibility of propositions or simultaneous (non-simultaneous) measurability because all propositions are compatible and there is no 'non-simultaneous measurability'. In quantum mechanics, however, this distinction does matter. We might then think of constructing a richer language say by introducing additional connectives (operators) reflecting non-simultaneous measurability (non-compatibility) of propositions into the language that would allow statements about compatible (incompatible) and simultaneously (non-simultaneously) measurable observables in quantum mechanics. There is no reason to believe that such statements should not 'conform' to classical logic, at least in the sense that the usual propositional connectives combining them should not behave classically. And, in fact, the (informal) language physicists use in reasoning about quantum systems is of such a nature, and the (propositional) logic they use is classical logic. Birkhoff and von Neumann were well aware of this, and it is important to note that this is exactly what they did not want in building 'the logic of quantum mechanics' as is obvious from John von Neumann's letter quoted above.

Phrased in modern terminology, Birkhoff and von Neumann want to retain the *language of propositional logic* as the language of the 'logic of quantum mechanics'. That is what they mean by a calculus that "resembles the usual calculus of propositions with respect to *and*, *or* and *not*". The novel logical notions mentioned are to be reflected in the propositional *(quantum) logic* to be constructed. This logic of course cannot be expected to be classical.

Note that the term 'calculus' is obviously not used in the modern sense. It becomes evident from the further progression of the paper that what the authors have in mind is not (necessarily) a deductive system. Rather it is reminiscent of algebraic logic where logics can be defined via algebraic structures. In algebraic logic, Boolean algebras for instance define classical logic. Generally, the paper does not display the distinction between syntax and semantics common in presenting logical systems in modern style.

In any case it seems that Birkhoff and von Neumann envisage a logical system in (whatever sense) or (in their terminology) logical structures into which notions such as non-commutativity (of observations) or non-simultaneous measurarability, i.e. notions which at first glance seem to have nothing to do with logic, can be incorporated. As indicated these novel features of quantum mechanics should be part of the *logic* rather than the *language of the logic*. Viewing the uncertainty relations as being logical in nature is a basic intuition of the Birkhoff-von Neumann paper. This is an insight which also plays a vital role in the approach to quantum logic taken in this book.

The quotations above are from the Introduction of the paper.

Let us now take a closer look at the proper contents of the paper.

3 Experimental propositions

In the first paragraphs of "Physical Background" the authors set out to explicate the concept of an *experimental proposition*. In this they build on various other concepts. First there is the concept of a *physical system*, which is unproblematic and taken for granted by the authors. Another basic concept is that of a *set of compatible measurements*, in modern terminology, a set of mutually compatible observables represented by mutually commuting Hermitian operators. They then proceed to the concept of an *observation* on $\mathcal{S}$: Let $\mu_1, ... \mu_n$ be n compatible measurements with outcomes $x_1, ..., x_n$. The observation amounts to specifying these values. Call the set of all n-tuples that can arise as values in compatible measurements an *observation space* of the system. Note that the concept of an observation space is relative to a finite number of compatible observations. So, in case $n = 1$, we have just one observable and the observation space corresponding to this observable is the set of all possible values it can assume as a result of an experiment. It is then natural to define an experimental proposition to be a subset of an observation space. We may view such an experimental proposition as a sort of prediction saying that the value of an observable that we get in a certain experiment belongs to a certain subset of the observation space. It is made clear, however, in the paper that not every subset of observation space is a proper candidate for this. It would for instance be absurd, as mentioned by Birkhoff and von Neumann, to call the assertion that the angular momentum of of the earth around the sun was at a particular instant a rational number an experimental proposition. What, however, is important to note is that in classical physics those subsets of observation space

3. EXPERIMENTAL PROPOSITIONS

that do represent experimental propositions must form a Boolean algebra with respect to the usual set operations. This reflects the requirement that classical mechanics 'conforms' to classical logic. Those subsets of observation space that actually represent experimental propositions have later been described as Borel measurable sets. The next crucial concept is that of a *phase space*. In classical mechanics, phase space means the following. Given a physical system of, say n particles. Then the 'state' of this system is characterised by the positions and momenta of these particles. Using the words of Birkhoff and von Neumann: " Thus, in classical mechanics, each point of Σ corresponds to a choice of n position and momentum coordinates... Hence in this case Σ is a region of $2n$-dimensional space."

What, now, is the analogue of this in quantum mechanics according to Birkhoff and von Neumann? They say: "Similarly, in quantum theory the points of Σ correspond to so-called wave functions, and hence Σ is again a function-space, usually assumed to be a Hilbert space."

Hilbert space as the phase space in quantum mechanics enters the stage here by way of analogy. The (heuristic) argument is this. The states of a classical system are determined by a tuple of positions and momenta and the phase space of the system is therefore the set of these tuples. In quantum mechanics the state of the system is determined by a wave function and therefore its phase space is the space of its wave functions, which is a Hilbert space.

3.1 A propositional calculus for quantum mechanics

From the conceptual point of view, the core of the first part of the paper is the paragraph 6 entitled "A propositional calculus for quantum mechanics", in which the core logical structure of the Birkhoff-von Neumann approach to quantum logic is introduced. It seems to us that this paragraph is not easy to read. We will therefore try to reconstruct the subtle and at times heuristic reasoning in detail so as to get a clear picture of what the authors precisely have in mind.

So far there is the concept of an experimental proposition defined as a subset of on observation space. What is needed now is to make the connection between experimental propositions and phase space in quantum mechanics, i.e. Hilbert space. Put differently, the question has to be answered which subsets of a Hilbert space (mathematically) represent experimental propositions. Nowadays, we are familiar with the following answer to this question. An experimental proposition is mathematically represented as a closed subspace of a Hilbert space. Let us see how Birkhoff and von Neumann arrive at this conclusion.

We quote: "The present section will be devoted to defining such a connection, proving some facts about it, and obtaining from it heuristically by introducing a plausible postulate, a propositional calculus for quantum mechanics".

Note that the authors consider the argument heuristic and that a 'plausible postulate' plays a role in it.

They continue: "Accordingly, let us observe that if $\alpha_1, ..., \alpha_n$ are any compatible observations on a quantum-mechanical system $\mathcal{S}$ with phase-space Σ, then there exists a set of mutually orthogonal closed linear subspaces Ω_i of Σ (which correspond to the families of proper functions satisfying $\alpha_1 f = \lambda_{i,1} f, ..., \alpha_n f = \lambda_{i,n} f$) such that *every* point (or function) $f \in \Sigma$ can be uniquely written in the form

$$f = c_1 f_1 + c_2 f_2 + c_3 f_3 + ... [f_i \in \Omega_i]"$$

Let us reconstruct this in modern terminology. The above sum is obviously 'infinite'. It is what we nowadays call the Fourier expansion of a vector in terms of a complete orthonormal system, see 2.13.

Here the term 'compatible observation' must be made precise mathematically. The context suggests that here $\alpha_1, ..., \alpha_n$ represent mutually commuting Hermitian operators. Then what they call the family of 'proper functions' is the family of eigenfunctions (eigenvectors) common to $\alpha_1, ..., \alpha_n$ and $\lambda_{i,1}, ..., \lambda_{i,n}$ are the corresponding eigenvalues. By "..such that..." the fact is expressed that the (normalised) eigenvectors of a Hermitian operator form a complete orthonormal system. In case $n = 1$ the above just says in modern terminology that any vector has a unique Fourier expansion in terms of eigenvectors of the Hermitian operator representing α_1

The text proceeds as follows: "Hence if we state the

DEFINITION 4.1. By the 'mathematical representative' of a subset S of any observation-space (determined by compatible observations $\alpha_1, ..., \alpha_n$) for a quantum- mechanical system $\mathcal{S}$, will be meant the set of all points f of the phase-space of $\mathcal{S}$, which are linearly determined by proper functions f_k satisfying $\alpha_1 f_k = \lambda_1 f_k, ..., \alpha_n f_k = \lambda_n f_k$, where $(\lambda_1, ..., \lambda_n) \in \mathcal{S}$)

Then it follows immediately: (1) that the mathematical representative of any experimental proposition is a closed linear subspace of Hilbert space (2) since all operators of quantum mechanics are Hermitian, that the mathematical representative of the *negative* of any experimental proposition is the *orthogonal complement* of the mathematical representation of the proposition itself..."

Note that they define the negative of an experimental proposition (or subset S of observation space) to be the experimental proposition corresponding to the set-complement of S in the same observation space.

3. EXPERIMENTAL PROPOSITIONS 49

Again, we think that this needs interpretation. Here, for the first time, the paper uses the term 'closed (linear) subspace'. Why closed subspace? What does 'linearly determined' mean? Assume 'linearly determined' means 'linearly spanned' in modern terminology. Then the conclusion that the resulting subspaces are closed would not be justified. Recall that in an infinite-dimensional Hilbert space the linear span of a set of vectors need not be closed although we do have this in the finite-dimensional case, see 3.

It seems that the only way of making sense of the above argument is this. Again let us assume the case $n = 1$, i.e. the observation space is determined by one observable. Let α denote the (Hermitian) operator representing this observable and let an experimental proposition $\mathcal{P}$ be given. Thus $\mathcal{P}$ is a subset of the set of eigenvalues of α. Let $\{x_\lambda \mid \lambda \in \mathcal{P}\}$ be the set of eigenvectors corresponding to the elements of $\mathcal{P}$. According to the above definition the mathematical representative of $\mathcal{P}$ is the of set of vectors which are 'linearly determined' in Birkhoff and von Neumann's terminology by the x'_λs. If, however, we take 'linearly determined' by 'linearly spanned' in modern terminology this span is not necessarily a closed subspace and conclusion (1) would be wrong since the linear span of a set of vectors of a Hilbert space need not be closed in the case of an infinite-dimensional vector space. This is generally true only if this subspace is finite-dimensional. By the term Hilbert space Birkhoff and von Neumann mean what in modern terminology is an infinite-dimensional Hilbert space. Therefore 'linearly determined' cannot mean 'linearly spanned' in modern terminology. We can, however, make perfect sense of the argument as follows. Take 'linearly determined as meaning to be a finite or infinite sum of the x_λ's, where infinite sum means a Fourier expansion in the x_λ's. Then the set of the "infinite sums" is the boundary of the span of the $x'_\lambda s$ and the resulting space is in fact a closed subspace. On this interpretation conclusion (2) is correct too.

For the sake of clarity let us put it this way. Consider the subspace spanned by the x_λ's. Then by definition any vector of this subspace is a linear combination of the x_λ's. Now take the (topological) closure of this subspace. This is a closed subspace and thus a Hilbert space itself. Any vector of this space has a a Fourier expansion in terms of the x_λ's. It seems to us that this is what Birkhoff and von Neumann mean by 'linearly determined'. The term means 'being a linear combination or the result of a Fourier expansion'.

To summarise, the following has been achieved. So far, four meaningful concepts have been introduced in the paper, namely the concepts of an observation space, an experimental proposition, the concept of a phase space, which in quantum mechanics is a Hilbert space, and the concept of the mathematical representation of an experimental proposition. The main conclusion of the partially heuristic but nevertheless convincing reasoning so far is that an experimental proposition should be mathematically represented by a closed subspace of a Hilbert space.

The paper proceeds by introducing a *Postulate*: "The set-theoretical product of any two mathematical representatives of experimental propositions concerning a quantum-mechanical system, is itself the mathematical representative of an experimental proposition"

What does this postulate say? Given any two closed subspaces A and B representing the experimental propositions P and Q. We can then consider $A \cap B$ which is again a closed subspace. The question is whether there is an experimental proposition having $A \cap B$ as its mathematical representative. It is important to note that this cannot be taken for granted. In order to understand the meaning of the *Postulate* recall the concept of an experimental proposition. An experimental proposition presupposes an observation space which in turn presupposes a (finite) set of compatible observables, in the simplest case one observable. Note that these are purely physical notions. What the *Postulate* says is this. Given two observation spaces $\mathcal{S}_1$ and $\mathcal{S}_2$ and two experimental propositions P_1 and P_2 respectively which are mathematically represented by the closed subspaces A and B respectively. Then it is *postulated* that there exists an observation space $\mathcal{S}_3$ and an experimental proposition P_3 relative to $\mathcal{S}_3$ such that the subspace $A \cap B$ is the mathematical representation of P_3. So the *Postulate* is about possible observation spaces and thus about possible observables. It may be viewed as a physical postulate or at least as a postulate concerning the link between physical observables and the logic and the formalism of quantum mechanics. It is in this light that the ensuing remark is to be understood: "This postulate would clearly be implied by the not unusual conjecture that all Hermitian-symmetric operators in Hilbert space (phase space) correspond to observables"

The reasoning now naturally proceeds as follows. Since the orthogonal complement of a closed subspace is a closed subspace and the linear sum and the set product of any two closed subspaces A and B are again closed subspaces, they conclude:

"The set product and closed linear sum of any two, and the orthogonal complement of any one closed linear subspace of Hilbert space representing mathematically an experimental proposition concerning a quantum-mechanical system $\mathcal{S}$, is itself the representation of an experimental proposition concerning $\mathcal{S}$"

They continue: "This defines the calculus of experimental propositions

3. EXPERIMENTAL PROPOSITIONS

concerning $\mathcal{S}$, as a calculus with three operations and a relation of implication..."

There is a passage in paragraph 4 which also needs interpretation. "In quantum theory ...the possibility of predicting in general the readings from measurements on a physical system $\mathcal{S}$ from a knowledge of its 'state' is denied; only statistical predictions are always possible. This has been interpreted as a renunciation of the doctrine of pre-determination; a thoughtful analysis shows that another and more subtle idea is involved. The central idea is that physical quantities are *related*, but are not all computable from a number of *independent basic* quantities (such as position and velocity). We will show in paragraph 12 that this situation has an exact algebraic analogue in the calculus of propositions."

What now is said in paragraph 12? Here they say: "...we conclude that *the propositional calculus of quantum mechanics has the same structure as an abstract projective geometry.*"

Let us try an interpretation of this statement. In classical mechanics all physical quantities can be computed, as is the terminology of Birkhoff and von Neumann, from certain basic quantities, namely position and velocity. If we substitute 'deducible' for 'computable' here, the above statement could mean that the 'calculus' of the propositions of quantum mechanics is not a deductive system (calculus) as is classical logic. Rather it has the structure of a (projective) geometry describing *relations* between states and propositions in analogy with the relations we have in a projective geometry between points, lines, hyper-planes. On this view the quantum mechanical calculus is geometrical rather than deductive.

We now give a short summary of part 2 of the paragraph entitled "Algebraic Analysis". As indicated in the title, this part is more technical in nature and less heuristic than the first part.

First, it is suggested that the calculus of experimental propositions should have an implication and this implication should be set inclusion.

The authors then proceed by defining the concept of a lattice. They suggest that the experimental propositions should form a lattice and should thus satisfy the 'laws' that hold in any lattice such as commutativity of join and meet and also associativity.

They then define the concept of a complemented lattice remarking that in the case of closed subspaces of a Hilbert space complementation is orthogonal complement formation.

In paragraph 10 an important issue is discussed, namely the question whether the calculus of quantum mechanics should satisfy the distributive identity, in Birkhoff-von Neumann notation

$$L6:\ a \cup (b \cap c) = (a \cup b) \cap (a \cup c)$$

as well as its dual form. It is argued that the calculus of quantum mechanics does not satisfy the distributive identity. This is even considered the "central difference" between the logic of classical and quantum mechanics. They write: "It is interesting that L6 is also a logical consequence of the compatibility of the observables occurring in a and b and c... These facts suggest that the distributive law *may* break down in quantum mechanics"

Note that this passage is somewhat vague. What does the phrase "compatible observables occurring in ..." mean? Again, it seems that this argument is heuristic. Birkhoff-von Neumann do not yet have the concept of compatibility of elements in a lattice. It is, however, suggested that a weakened form of the distributive law should hold, namely the following identity called *modularity*, see 2.16, in Birkhoff-von Neumann notation

$$L5 \text{ If } a \subset c, \text{ then } a \cup (b \cap c) = (a \cup b) \cap c$$

It is pointed out that finite-dimensional subspaces of a Hilbert space do satisfy the modular identity. Moreover, it is shown by a counterexample that this is not the case for infinite-dimensional closed subspaces.

Then an interesting mathematical observation is made, namely that the modular identity follows from the existence of a 'numerical dimension function' d, i.e. a function with the following properties.

$$D1: \text{ If } a > b, \text{ then } d(a) > d(b)$$

and

$$D2: d(a) + d(b) = d(a \cap b) + d(a \cup b)$$

Note that the modular identity is stronger than the orthomodular identity. It is interesting that Birkhoff-von Neumann insist on this stronger version which, as already noted, does not necessarily hold for infinite-dimensional closed subspaces of a Hilbert space. As pointed out by Redei in [52] it was von Neumann's hope that such a dimension function, which is similar to a probability function, might be of use in understanding the probabilistic nature of quantum mechanics.

What was Birkhoff and von Neumann's attitude towards their paper, which was to turn out seminal later on?

A letter by von Neumann to Birkhoff written before the paper was published may give us some insight. It contains the following passage (from [53]): "Your general remarks, I think, are very true: I too think that our paper will not be exhaustive or conclusive, but that we should not attempt to make it such : The subject is obviously only at the beginning of a development, and we want to suggest the direction of this development much more, than reach 'final results'. I, for one, do not even believe, that the right formal frame for quantum mechanics is already found."

3.2 The Kochen-Specker and the Schütte tautologies

Although Birkhoff and von Neumann must have been fully aware of the fact that their logical 'calculus' must differ from classical logic in important respects, they did not provide an in depth comparison between the two logics. The only difference they explicitly mention is that the distributive laws of classical logic no longer hold in the 'logic of quantum mechanics'. In view of the fact that it was their chief motivation to incorporate the profound differences between classical and quantum mechanics such as the existence of uncertainty relations in quantum mechanics into their logic this seems not too profound an observation. It took some time until a truly surprising phenomenon was discovered which highlights the profound difference between Birkhoff-von Neumann quantum logic and classical logic. Namely, Kochen and Specker discovered in their classic paper "The problem of hidden variables in quantum mechanics" [34], as a byproduct, the following fact. There exists a classical tautology which under a certain 'valuation' in the lattice of closed subspaces in (three-dimensional) Hilbert space represents the zero space. This means that there exists a classical tautology which is a 'quantum logical contradiction' and the other way round. Kochen and Specker explicitly present such a tautology in 117 propositional variables. A similar tautology had even before the publication of the Kochen-Specker paper been found by Schütte as is known from a letter that Schütte wrote to Specker (see [10]). Schütte's tautology does not represent the zero space. But it does not represent the whole space either. It is a classical tautology which is not a quantum tautology. It is important to note that both in the Kochen-Specker tautology and in Schütte's tautology only compatible quantum propositions are combined via the connectives.

We will study this phenomenon of classical inconsistency in the Birkhoff-von Neumann quantum logic from a general point of view in chapters 6 and 8. We will see that this phenomenon is by no means accidental.

CHAPTER 5

THE DYNAMIC VIEWPOINT: PROPOSITIONS AS OPERATORS

1 Propositions viewed dynamically

Let us begin by pointing out a certain analogy between measurements and propositions. A physical measurement, e.g., measuring the temperature of a gas to be $138\,°K$, asserts that the proposition *the temperature of this gas is* $138\,°K$ holds true. A measurement, in a sense, asserts the truth of a proposition. This is the fundamental analogy between physics and logic: making a measurement is similar to asserting a certain kind of proposition. The example above has been taken from classical physics. Consider now measuring the spin of a particle along the z-axis to be $1/2$. This measurement is akin to asserting the truth of the proposition *the spin along the z-axis is* $1/2$. But, here, the assertion of the proposition, i.e., the measurement, changes the state of the system. The assertion holds in the state resulting from the measurement, but did not necessarily hold in the state of the system before the measurement was performed. In fact it held in this previous state if and only if the measurement left the state unchanged. Inspired by the analogy between measurements and propositions we set ourselves to study the logic of propositions that not only *hold* at states but also *act* on them, transforming the state in which they are evaluated into another one. A proposition holds in some state if and only if this state is a fixed point for the proposition.

Conceptually, this is the novelty of our approach to logic in this chapter. We view propositions in a dynamic rather than in a static way. In terms of Hilbert space this means that we view propositions as projections on closed subspaces rather than closed subspaces.

We will see that in this dynamic framework classical logic appears as the static limiting case.

2 The concept of an M-Algebra

Inspired by the above consideration we define a class of abstract structures for which we coin the term *algebras of measurements*, M-algebras for short.

Formally, this concept is an abstraction from the algebra of projections of a Hilbert space.

The structures we are concerned with deal with a set X and a set M of functions from X to X. We think of X as a set of states and we think of M as a set of propositions (measurements) acting on these states.

We will denote the composition of functions by $\circ$ and composition has to be understood from left to right: for any $x \in X$, $(\alpha \circ \beta)(x) = \beta(\alpha(x))$. If $\alpha : X \to X$, we will denote by $FP(\alpha)$ the set of all fixed points of α: $FP(\alpha) \stackrel{\text{def}}{=} \{x \in X \mid \alpha(x) = x\}$.

DEFINITION 5.1. An M-algebra is a pair $\langle X, M \rangle$ in which X is a non-empty set and M is a set of functions from X to X, that satisfies the six properties described below.

- **Illegitimate** $\exists\, 0 \in X$ such that $\forall \alpha \in M$, $0 \in FP(\alpha)$, i.e, $\alpha(0) = 0$.
- **Idempotence** $\forall \alpha \in M$, $\alpha \circ \alpha = \alpha$, i.e., for any $x \in X$, $\alpha(\alpha(x)) = \alpha(x)$.

  The next property requires a preliminary definition.

  DEFINITION 5.2. For any $\alpha, \beta : X \to X$, we will say that α *preserves* β if and only if α preserves $FP(\beta)$, i.e., if $\alpha(FP(\beta)) \subset FP(\beta)$, i.e., $\forall x \in X$, $\beta(x) = x \Rightarrow \beta(\alpha(x)) = \alpha(x)$.

- **Composition** $\forall \alpha, \beta \in M$, if α preserves β, then $\beta \circ \alpha \in M$.
- **Interference** $\forall x \in X$, $\forall \alpha, \beta \in M$, if $x \in FP(\alpha)$, i.e., $\alpha(x) = x$, and $(\beta \circ \alpha)(x) \in FP(\beta)$, i.e., $\beta(\alpha(\beta(x))) = \alpha(\beta(x))$, then $\beta(x) \in FP(\alpha)$, i.e., $\alpha(\beta(x)) = \beta(x)$.
- **Cumulativity** $\forall x \in X$, $\forall \alpha, \beta \in M$, if $\alpha(x) \in FP(\beta)$, i.e., $\beta(\alpha(x)) = \alpha(x)$ and $\beta(x) \in FP(\alpha)$, i.e., $\alpha(\beta(x)) = \beta(x)$, then $\alpha(x) = \beta(x)$.

  The next property requires some notation. For any $\alpha : X \to X$, we will denote by $Z(\alpha)$ the set of zeros of α: $Z(\alpha) \stackrel{\text{def}}{=} \{x \in X \mid \alpha(x) = 0\}$.

- **Negation** $\forall \alpha \in M$, $\exists (\neg \alpha) \in M$, such that $FP(\neg \alpha) = Z(\alpha)$, and $Z(\neg \alpha) = FP(\alpha)$, i.e., $\forall x \in X$, $\alpha(x) = 0$ iff $(\neg \alpha)(x) = x$ and $\forall x \in X$, $\alpha(x) = x$ iff $(\neg \alpha)(x) = 0$.

An additional property will be considered in Section 3.11.

DEFINITION 5.3. An M-algebra is *separable* if it satisfies the following:

Separability For any $x, y \in X - \{0\}$, if $x \neq y$ then $\exists \alpha \in M$ such that $\alpha(x) = x$ and $\alpha(y) \neq y$.

3. MOTIVATION AND JUSTIFICATION 57

The definition of an M-algebra needs comment. The mathematically educated reader will easily realise that the structures called M-algebras defined above do, in the strict sense of Universal Algebra, not qualify as algebras. From the point of view of Universal Algebra as well as from the viewpoint of Model Theory, algebras are first-order structures whereas what we call an M-algebra is a second order structure very much akin to a topological space. The elements of M are second order entities as are the open sets of a topological space. The axioms of both M-algebras and a topological spaces are formulated in a second order language quantifying over second order entities. We retain the term *M-algebra* for the sake of a suggestive terminology reminding the reader of the analogy with measurements in quantum mechanics.

3 Motivation and justification

The reader may think of the defining properties of M-algebras as playing the role of axioms in classical logic. In the static limiting case of classical logic we have to axiomatise the truth of propositions. In the general dynamic case we need to axiomatise the action of propositions.

In this section, we will leisurely explain each one of these properties. Our explanation of each property will include three parts:

- an epistemological explanation whose purpose is to explain why the property is natural or even required when one thinks of measurements,

- an explanation of why the property holds in the algebra $\langle H, L \rangle$ where H is a Hilbert space and L the set of all projections onto closed subspaces of H,

- an explanation of the logical meaning of the property, based on the identification of measurements with propositions.

3.1 States

In the Hilbertian description of quantum mechanics, a (pure) state is a one-dimensional subspace, i.e., a ray, in some Hilbert space. The illegitimate state, 0 is the zero-dimensional subspace.

In our study of M-algebras a state is a primitive notion and we need not reflect in depth on this at this stage. We will enter a detailed discussion of the concept of a state in chapters 6 and 8.

3.2 Measurements

We may think of the elements of M as representing measurements on the physical system whose possible states are those of X. In classical physics one may assume that a measurement leaves the measured system unchanged. It is a hallmark of quantum mechanics that this assumption cannot be held true anymore. In quantum mechanics, measurements, in general, change the state of the system. This is the phenomenon called *collapse of the wave function*. Therefore we model measurements by transformations on the set of states. Clearly not any transformation can be called a measurement. A measurement changes the system in some minimal way. A transformation that brings about a wild change in the system cannot be considered to be a measurement.

In the Hilbertian description of quantum mechanics measurable quantities are represented by Hermitian operators. Measurements in our sense are represented by a pair $\langle A, \lambda \rangle$ where A is a Hermitian operator and λ an eigenvalue of A. The effect of measuring $\langle A, \lambda \rangle$ in state x is to project x onto the eigensubspace of A for eigenvalue λ. A measurement α is therefore a projection on a closed subspace of a Hilbert space. The set $FP(\alpha)$ is the closed subspace on which α projects. Those projections onto eigensubspaces are the measurements we try to identify. Our goal is to identify the algebraic properties of such projections that make them suitable to represent physical measurements in quantum mechanics.

This is the interpretation that we will take along with us: a measurement α *holds* at some state x, or, equivalently x satisfies α, if and only if $x \in FP(\alpha)$.

3.3 Illegitimate

Illegitimate is mainly a technical requirement. The sequel will show why it is handy. The illegitimate state 0 is a state that is physically impossible. Physicists, in general, do not consider this state explicitly. From the epistemological point of view, we just require that amongst all the possible states of the system we include a state, denoted 0 that represents physical impossibility. There is not much sense in measuring anything in the illegitimate state, therefore, it is natural to assume that no measurement α operating on the illegitimate state can change it into some legitimate state. This is the meaning of our requirement that 0 be a fixed point of any measurement. In other terms, the state 0 satisfies every measurement, every measurement holds at 0.

In the Hilbertian description of quantum mechanics the zero vector plays the role of our 0. Indeed, since a projection is linear, it preserves the zero vector.

From a logician's point of view **Illegitimate** requires us to include the inconsistent theory in X. Clearly, the result of adding any proposition to the inconsistent theory leaves us with the inconsistent theory.

3.4 Zeros

We have described in Section 3.2 the interpretation we give to the fact that a state x is a fixed point of a measurement α. We want to give a similarly central meaning to the fact that a state x is a zero of a measurement α: $x \in Z(\alpha)$, i.e., $\alpha(x) = 0$. If measuring α sends x to the illegitimate state, measuring α is physically impossible at x. This should be understood as meaning that the state x has some definite value different from the one specified by α.

If, at x, the spin is $1/2$ along the z-axis, then measuring along the z-axis a spin of $-1/2$ is physically impossible and therefore the measurement of $-1/2$ sends the state x to the illegitimate state 0. The status of the measurement that measures $-1/2$ along the *x-axis* is completely different: this measurement does not set x to 0, but to some legitimate state in which the spin along the x-axis is $-1/2$.

It is natural to say that a measurement α has a definite value at x iff x is either a fixed point or a zero of α. We will define: $Def(\alpha) \stackrel{\text{def}}{=} FP(\alpha) \cup Z(\alpha)$. If $x \in Def(\alpha)$, α has a definite value at x: either it holds at x or it is impossible at x. If $x \notin Def(\alpha)$, $\alpha(x)$ is some state different from x and different from 0.

In the Hilbertian presentation of quantum mechanics, the zeros of a measurement α are the vectors orthogonal to the set of fixed points of α.

3.5 Idempotence

Idempotence is extremely meaningful. It is an epistemologically fundamental property of measurements that they are idempotent: if α is a measurement and x a state, then $\alpha(\alpha(x)) = \alpha(x)$, i.e., measuring the same value twice in a row is exactly like measuring it once. Note that, by **Illegitimate**, if $x \in Def(\alpha)$, then $\alpha(\alpha(x)) = \alpha(x)$. The importance of **Idempotence** concerns states that are not in $Def(\alpha)$.

It seems very difficult to imagine a scientific theory in which measurements are not idempotent: it would be impossible to check directly that a system is indeed in the state we expect it to be in without changing it. Idempotence is one of the conditions that ensure that measurements change states only minimally. This principle seems to be a fundamental principle of all science, having to do with the reproducibility of experiments. If there was a physical system and a measurement that, if performed twice in a row gave different results, then such a measurement would be, in principle, irreproducible.

In the Hilbertian description of quantum mechanics measurements are modelled by projections onto eigensubspaces. Any projection is idempotent. But it is enlightening to reflect on the phenomenology of this idempotence.

For an electron whose spin is positive along the z-axis (state x_0), measuring a negative spin along the x-axis is feasible, i.e., does not send the system into the illegitimate state, but sends the system into a state x_1 different from the original one, x_0. Nevertheless, a consequence of the collapse of the wave function is that, after measuring a negative spin along the x-axis, the spin is indeed negative along the x-axis and therefore a new measurement of a negative spin along the x-axis leaves the state x_1 of our electron unchanged, whereas measuring a positive spin along the x-axis is now an unfeasible measurement and sends x_1 to the illegitimate state. Note that such a measurement of a positive spin along the x-axis in the original state x_0 brings us to a legitimate state x_3 different from x_0 and x_1. The idempotence of measurements, probably epistemologically necessary, provides some explanation of why projections in Hilbert spaces are a suitable model.

From the logical point of view, idempotence corresponds to the fact that asserting the truth of a proposition is equivalent to asserting it twice.

3.6 Preservation

The definition of *preservation* encapsulates the way in which different measurements can interfere. If α preserves $FP(\beta)$, the set of states in which β holds, α never destroys the truth of proposition β: it never interferes badly with β.

3.7 Composition

Composition has physical significance. It is a global principle: it assumes a global property and concludes a global property. Measurements are mappings of X into itself, therefore we may consider the composition of two measurements. According to the principle of minimal change, we do not expect the composition of two measurements to be a measurement: two small changes may make a big change. But, if those two measurements do not interfere in any negative way with each other, we may consider their composition as small changes that do not add up to a big change. **Composition** requires that if, indeed, α preserves β, then the composite operation that consists of measuring β first, and then α does not add up to a big change and should be a bona fide measurement. Notice that we perform β first, whose result is (by **Idempotence**) a state that satisfies β, then we perform α, which does not destroy the result obtained by the first measurement β.

In the Hilbertian presentation of quantum mechanics, consider α, the projection on some closed subspace A and β, the projection on B. The measurement α preserves β iff the projection of the subspace B onto A is contained in the intersection $A \cap B$ of A and B. In such a case the composition $\beta \circ \alpha$ of the two projections, first on B and then on A is equivalent to the projection on the intersection $A \cap B$. It is therefore a projection on

3. MOTIVATION AND JUSTIFICATION

some closed subspace.

Technically, the role of **Composition** is to ensure that the composition of two commuting measurements is a measurement. Equivalently, we could have, instead of **Composition**, required that for any pair $\alpha, \beta \in M$ such that $\alpha \circ \beta = \beta \circ \alpha$, their composition $\alpha \circ \beta$ be in M.

3.8 Interference

Interference has a deep physical meaning. It is a local principle, i.e., holds separately at each state x. It may be seen as a local logical version of Heisenberg's uncertainty principle. It considers a state x that satisfies α. Measuring β at x may leave α undisturbed (this is the conclusion), but, if β disturbs α, then no state at which both α and β hold can ever be attained by measuring α and β in succession. In other words, either such a state, satisfying both α and β is obtained immediately, or never.

We will say that β disturbs α at x if $x \in FP(\alpha)$ but $\beta(x) \notin FP(\alpha)$. Note that β preserves α if and only if it disturbs α at no x. **Interference** says that if β disturbs α at x then α disturbs β at $\beta(x)$, and β disturbs α at $(\beta \circ \alpha)(x)$, and so on. We chose to name this property *Interference* since it deals with the local interference of two measurements: if they interfere once, they will continue interfering ad infinitum.

In the Hilbertian presentation of quantum mechanics, the principle of **Interference** is satisfied for the following reason. Consider a vector $x \in H$ and two closed subspaces of H: A and B. Assume x is in A. Let y be the projection of x onto B and z the projection of y onto A. Assume that z is in B. Since both x and z are in A, the vector $z - x$ is in A. Similarly, the vector $z - y$ is in B. But y is the projection of x onto B and therefore $y - x$ is orthogonal to B and in particular orthogonal to $z - y$. We have $\langle y - x, z - y \rangle = 0$, and

$$\langle y, z \rangle - \langle y, y \rangle - \langle x, z \rangle + \langle x, y \rangle = 0.$$

Since z is the projection of y onto A, the vector $z - y$ is orthogonal to A and we have $\langle z - x, z - y \rangle = 0$, and

$$\langle z, z \rangle - \langle z, y \rangle - \langle x, z \rangle + \langle x, y \rangle = 0.$$

By subtracting the first equality from the second we get:

$$-\langle z, y \rangle - \langle y, z \rangle + \langle y, y \rangle + \langle z, z \rangle = \langle y - z, y - z \rangle = 0.$$

We conclude that $y = z$.

3.9 Cumulativity

Cumulativity is motivated by logic. It does not seem to have been reflected upon by physicists. It parallels the cumulativity property that is central to nonmonotonic logic: see for example [35]. If the measurement of α at x causes β to hold (at $\alpha(x)$), and the measurement of β at x causes α to hold (at $\beta(x)$) then those two measurements have, locally (at x), the same effect. Indeed, they cannot be directly distinguished by testing α and β. **Cumulativity** says that they cannot be distinguished even indirectly.

In the Hilbertian formalism, if the projection, y, of x onto some closed subspace A is in B (closed subspace) then y is the projection of x onto the intersection $A \cap B$. If the projection z of x onto B is in A, z is the projection of x onto the intersection $B \cap A$ and therefore $y = z$. In fact, a stronger property than **Cumulativity** holds in Hilbert spaces. The following property, similar to the Loop property of [35], holds in Hilbert spaces: **L-Cumulativity** $\forall x \in X$, for any natural number n and for any sequence $\alpha_i \in M$, $i = 0, \ldots, n$ if, for any such i, $\alpha_i(x) \in FP(\alpha_{i+1})$, where $n+1$ is understood as 0, then, for any $0 \leq i,j \leq n$, $\alpha_i(x) = \alpha_j(x)$.

To see that this property holds in Hilbert spaces, consider the distance d_i between x and the closed subspace A_i on which α_i projects. The condition $\alpha_i(x) \in FP(\alpha_{i+1})$ implies that $d_{i+1} \leq d_i$. We have $d_0 \geq d_1 \geq \ldots \geq d_n \geq d_0$ and we conclude that all those distances are equal and therefore $\alpha_i(x) \in FP(\alpha_{i+1})$ implies that $\alpha_i(x) = \alpha_{i+1}(x)$. We do not know whether the stronger **L-Cumulativity** is meaningful for quantum mechanics, or simply an uninteresting consequence of the Hilbertian formalism.

3.10 Negation

Negation also originates in logic. It corresponds to the assumption that propositions are closed under negation. If α is a measurement, α tests whether a certain physical quantity has a specific value μ. If such a test can be performed, it seems that a similar test could be performed to test the fact that the physical quantity of interest has some other specific value or does not have value μ.

In the Hilbertian formalism, to any closed subspace corresponds its orthogonal subspace, also closed.

3.11 Separability

We remind the reader that **Separability** is not included in the defining properties of an M-algebra. **Separability** asserts that if any two non-zero states x and y are different, there is a measurement that holds at x and not at y. Indeed, if all measurements that hold at x also hold at y it would not be possible to be sure that the system is in x and not in y. Compared to the previous requirements, **Separability** is of quite a different kind. It is some akin to a superselection principle, though presented in a dual way: a restriction on the set of states not on the set of observables.

Note that this implies that, in any non-trivial M-algebra (an M-algebra is trivial if $X = \{0\}$ and $M = \emptyset$), every state satisfies some measurement.

In the Hilbertian formalism, the projections on the one-dimensional subspaces defined by x and y respectively do the job.

For the logician, if T_1 and T_2 are two maximal consistent sets that are different, there is a formula α in $T_1 - T_2$. But, one may easily find (non-maximal) different theories T_1 and T_2 such that $T_1 \subset T_2$, contradicting **Separability**.

4 Examples of M-algebras

In this section we will formally define paradigmatical examples of M-algebras.

4.1 Examples from classical logic

Let $\mathcal{L}$ be the language of propositional logic as introduced in chapter 1. Recall that we call $T \subset \mathcal{L}$ a *theory* iff it is closed under the (classical) consequence operation $\mathcal{C}n$, i.e. if $\mathcal{C}n(T) = T$. $\mathcal{C}n$ has the properties listed in 2.3 and is, in fact, characterised by those properties.

Let X be the set of all theories. We define the action of a formula $\alpha \in \mathcal{L}$ on a theory T by: $\alpha(T) = \mathcal{C}n(T \cup \{\alpha\})$. Note that $\alpha(T)$ is in fact a theory by the idempotence of $\mathcal{C}n$. Let M be the set of all mappings of the above form. Then $\langle X, M \rangle$ is an M-algebra. We denote the mapping induced by α itself by α.

In such a structure α holds at T iff $\alpha \in T$. This is seen as follows. Suppose $\alpha(T) = T$. This means that $\mathcal{C}n(T \cup \{\alpha\}) = T$, for which we also write $\mathcal{C}n(T, \alpha)$. Since $\alpha \in \mathcal{C}n(T, \alpha)$, we have $\alpha \in T$. For the other direction, if $\alpha \in T$, then clearly $\alpha \in \mathcal{C}n(T, \alpha)$. Also note that such structures are monotonic in the sense that $T \subset \alpha(T)$.

These structures satisfy all the defining properties of an M-algebra. The illegitimate state is the theory $\mathcal{L}$. **Idempotence** follows from the property of idempotence of the consequence operation $\mathcal{C}n$, see 2.3. **Composition** follows from **Conjunction**: the composition $\alpha \circ \beta$ is the measurement $\alpha \wedge \beta$. Note that any pair of measurements commute. **Interference** is satisfied because $\alpha \in T$ implies $\alpha \in \mathcal{C}n(T, \beta)$. **Cumulativity** is satisfied because $\beta \in \mathcal{C}n(T, \alpha)$ implies $\mathcal{C}n(T, \alpha) = \mathcal{C}n(T, \alpha, \beta)$ by **Monotonicity** and **Idempotence**. **Negation** holds by the property of $\mathcal{C}n$ of the same name.

This M-algebra does not satisfy **Separability** since there are theories T and S such that $T \subset S$ and every formula α satisfied by T is also satisfied by S. This M-algebra is *commutative*: any two measurements commute since $\mathcal{C}n(\mathcal{C}n(T, \alpha), \beta) = \mathcal{C}n(\mathcal{C}n(T, \beta), \alpha)$.

If we consider the subset $Y \subset X$ consisting only of *maximal consistent* theories and the inconsistent theory, we see that the pair $\langle Y, \mathcal{L} \rangle$ is an M-algebra using the same arguments as above. In this M-algebra, all measurements do more than commute, they are *classical*, in the following sense.

DEFINITION 5.4. A mapping $\alpha : X \to X$ is said to be *classical* iff for every $x \in X$, either $\alpha(x) = x$ or $\alpha(x) = 0$.

The M-algebra above is separable: if T_1 and T_2 are different maximal consistent theories there is a formula $\alpha \in T_1 - T_2$. Let us now require such an M-algebra to have the property that any non-zero state Σ has a characteristic measurement in the following sense. For any non-zero state Σ there exists a unique measurement σ such that $\sigma(\Sigma) = \Sigma$. In this case we even have a strong form of separability. We will call such M-algebras strongly separable in the next chapter. The states of such an M-algebra are reminiscent of the states of a classical physical system, i.e the set of physical statements true of the system. The requirement that there be a characteristic measurement for any (consistent) state reflects the fact that the state of a classical physical system can be characterised by a single statement, namely by the statement specifying the positions and momenta of all particles constituting the system. This state space of a classical system is in physics called phase space.

Another natural idea would be to consider revision systems a la AGM [1] as possible candidates for examples of M-algebras. The action of a formula α on a theory T would be defined as the theory T revised by α: $T*\alpha$. Let us, however, mention that the structure obtained does not satisfy the M-algebra properties. The most blatant violation concerns **Negation**. In traditional revision theory negation does not behave at all as expected in an M-algebra.

4.2 Orthomodular spaces

Given an orthomodular space H, let M be $PRO(H)$, i.e. the set of projections of of H. Then the pair $\langle H, M \rangle$ is an M-algebra. In fact, the arguments in the case of Hilbert spaces used in section 3 are also valid in the more general situation of orthomodular spaces. This M-algebra is not separable, though: any two colinear vectors satisfy the same measurements.

We get a separable M-algebra if we take X to be the set of one-dimensional or zero-dimensional subspaces of H. For this note that the projections of H act in a natural way on X. Let M be the set of the mappings induced by the projections on X. Then the pair $\langle X, M \rangle$ is easily seen to be an M-algebra. This M-algebra is separable. Namely, for any $x \in X$, x is the only state satisfying the measurement represented by the projection on x.

5 Properties of M-algebras

We assume that $\langle X, M \rangle$ is an arbitrary M-algebra. First, we will show that any M-algebra includes two trivial measurements: $\top$, analogous to the truth-value *true*, that leaves every state unchanged and measures a property satisfied by every state and $\bot$, analogous to *false*, that sends every state to the illegitimate state, and is nowhere satisfied.

LEMMA 5.5. *[Negation, Composition, Idempotence] There are measurements* $\top, \bot \in M$ *such that for every* $x \in X$, $\top(x) = x$ *and* $\bot(x) = 0$.

Proof. The set M of measurements is not empty: assume $\alpha \in M$. Clearly, by **Negation**, the measurement $\neg \alpha$ preserves α. It follows, by **Composition**, that $\alpha \circ (\neg \alpha)$ is a measurement. Let $\bot = \alpha \circ (\neg \alpha)$. By **Idempotence** and **Negation**, for every $x \in X$, $\bot(x) = 0$. We now let $\top = \neg \bot$. ∎

Then, we want to show that measurements are uniquely specified by their fixed points.

LEMMA 5.6 (Idempotence, Cumulativity). *For any* $\alpha, \beta \in M$, *if* $FP(\alpha) = FP(\beta)$, *then* $\alpha = \beta$.

Proof. Assume $FP(\alpha) = FP(\beta)$. Let $x \in X$. By **Idempotence** $\alpha(x) \in FP(\alpha)$ and therefore, by assumption $\alpha(x) \in \text{box} FP(\beta)$. Similarly $\beta(x) \in FP(\alpha)$. By **Cumulativity**, then, $\alpha = \beta$. ∎

COROLLARY 5.7 (Idempotence, Cumulativity, Negation). *For any* $\alpha \in M$, $\neg\neg\alpha = \alpha$.

Proof. Both α and $\neg\neg\alpha$ are measurements and $FP(\neg\neg\alpha) = FP(\alpha)$. ∎

We will now prove a very important property. Suppose x is a state in which some measurement (i.e., proposition) holds: for example, at x the spin along the x-axis is $1/2$. Performing a measurement α on x may lead to a different state $y = \alpha(x)$. At y the spin along the x-axis may still be $1/2$, or it may be the case that the measurement α has interfered with the value of the spin. But, under no circumstance, can it be the case that the spin along the x-axis has a definite value different from $1/2$, such as $-1/2$. If the value of the spin along the x-axis at y is not $1/2$, the spin must be indefinite. This expresses the fact that a measurement α, acting on a state in which β holds, can either preserve β (when $\alpha(x) \in FP(\beta)$) or can disturb β (when $\alpha(x) \notin Def(\beta)$) but cannot make β impossible at x, i.e., $\alpha(x) \in Z(\beta)$. This is a very natural requirement stemming from the *minimal change* principle. A move from a definite value to a different definite value is too drastic to be accepted as measurement.

In the Hilbertian presentation of quantum mechanics, measurements are projections. The projection of a non-null vector x onto a closed subspace A is never orthogonal to x, unless x is orthogonal to A. Therefore if x is in some subspace B, but its projection on A is orthogonal to B, then this projection is the null vector.

LEMMA 5.8 (Illegitimate, Interference). *For any $x \in X$, $\alpha, \beta \in M$, if $x \in FP(\beta)$, i.e., $\beta(x) = x$, and $\alpha(x) \in Z(\beta)$, i.e., $\beta(\alpha(x)) = 0$, then $x \in Z(\alpha)$, i.e., $\alpha(x) = 0$.*

Proof. Assume $x \in FP(\beta)$ and $\beta(\alpha(x)) = 0$. Then $(\alpha \circ \beta)(x) = 0 \in FP(\alpha)$. By **Interference**, then, $\alpha(x) \in FP(\beta)$ and $\beta(\alpha(x)) = \alpha(x)$, i.e., $0 = \alpha(x)$. ∎

We will now sort out the relation between fixed points and zeros. The next result is a dual of lemma 5.8.

LEMMA 5.9 (Illegitimate, Interference, Negation). *$\forall x \in X$, $\forall \alpha, \beta \in M$, if $x \in Z(\beta)$ and $\alpha(x) \in FP(\beta)$, then $x \in Z(\alpha)$. In other terms, if $\beta(x) = 0$ and $\beta(\alpha(x)) = \alpha(x)$, then $\alpha(x) = 0$.*

Proof. Consider the measurement $\neg \beta$ guaranteed by **Negation**. If we have $x \in FP(\neg \beta)$ and $\alpha(x) \in Z(\neg \beta)$, then, by lemma 5.8 we have $x \in Z(\alpha)$. ∎

LEMMA 5.10 (Illegitimate, Idempotence, Interference, Negation). *For any $\alpha, \beta \in M$, $FP(\alpha) \subset FP(\beta)$ iff $Z(\beta) \subset Z(\alpha)$.*

Proof. Suppose $FP(\alpha) \subset FP(\beta)$ and $x \in Z(\beta)$. Since, by **Idempotence**, $\alpha(x) \in FP(\alpha)$, we have, by assumption, $\alpha(x) \in FP(\beta)$. By lemma 5.9, then $x \in Z(\alpha)$.

Suppose now that $Z(\beta) \subset Z(\alpha)$. We have $FP(\neg\beta) \subset FP(\neg\alpha)$ and by what we just proved: $Z(\neg\alpha) \subset Z(\neg\beta$. We conclude that $FP(\alpha) \subset FP(\beta)$. ∎

We will now consider the composition of measurements. First we show the symmetry of the preservation relation.

LEMMA 5.11 (Idempotence, Interference). *For any $\alpha, \beta \in M$, α preserves β iff β preserves α.*

Proof. Assume α preserves β, and $x \in FP(\alpha)$. By **Idempotence**, $\beta(x) \in FP(\beta)$. Since α preserves β, $\alpha(\beta(x)) \in FP(\beta)$. The assumptions of **Interference** are satisfied and we conclude that $\beta(x) \in FP(\alpha)$. We have shown that β preserves α. ∎

5. PROPERTIES OF M-ALGEBRAS

LEMMA 5.12 (Illegitimate, Idempotence, Interference, Negation). *For any $\alpha, \beta \in M$, if $\alpha \circ \beta \in M$, then $FP(\alpha \circ \beta) = FP(\alpha) \cap FP(\beta)$.*

Proof. Since $Z(\alpha) \subset Z(\alpha \circ \beta)$, lemma 5.10 implies that $FP(\alpha \circ \beta) \subset FP(\alpha)$. By **Idempotence** of β, $FP(\alpha \circ \beta) \subset FP(\beta)$. We see that $FP(\alpha \circ \beta) \subset FP(\alpha) \cap FP(\beta)$. But the inclusion in the other direction is obvious. ∎

We will now show that the converse of **Composition** holds.

LEMMA 5.13 (Illegitimate, Idempotence, Interference, Negation). *For any $\alpha, \beta \in M$, if $\alpha \circ \beta \in M$, then β preserves α.*

Proof. By lemma 5.12, $FP(\alpha \circ \beta) \subset FP(\alpha)$. For any x, $(\alpha \circ \beta)(x)$ is therefore a fixed point of α. Assume $x \in FP(\alpha)$. Then, $(\alpha \circ \beta)(x) = \beta(x)$ is a fixed point of α. ∎

LEMMA 5.14 (Illegitimate, Idempotence, Interference, Composition, Negation). *For any $\alpha, \beta \in M$, $\alpha \circ \beta \in M$, iff β preserves α.*

Proof. The *only if* part is lemma 5.13. The *if* part is **Composition**. ∎

LEMMA 5.15 (Illegitimate, Idempotence, Interference, Composition, Negation). *For any $\alpha, \beta \in M$, $\alpha \circ \beta \in M$ iff $\beta \circ \alpha \in M$.*

Proof. By lemmas 5.14 and 5.11. ∎

LEMMA 5.16 (Illegitimate, Idempotence, Interference, Composition, Cumulativity, Negation). *For any $\alpha, \beta \in M$, $\alpha \circ \beta \in M$ iff α and β commute, i.e., $\alpha \circ \beta = \beta \circ \alpha$.*

Proof. Assume, first, that $\alpha \circ \beta \in M$. By lemma 5.15, $\beta \circ \alpha \in M$. By lemma 5.12, $FP(\alpha \circ \beta) = FP(\beta \circ \alpha)$, which implies the claim by lemma 5.6.

Assume, now that α and β commute. We claim that α preserves β: indeed, if $\beta(x) = x$, then $\beta(\alpha(x)) = \alpha(\beta(x)) = \alpha(x)$ and therefore, by **Composition**, $\beta \circ \alpha$ is a measurement. ∎

LEMMA 5.17 (Illegitimate, Idempotence, Interference, Composition, Cumulativity, Negation). *For any $\alpha, \beta \in M$, if $FP(\alpha) \subset FP(\beta)$, then $\alpha \circ \beta = \beta \circ \alpha = \alpha$.*

Proof. If $FP(\alpha) \subset FP(\beta)$, then, clearly $\alpha \circ \beta = \alpha$ by Idempotence of α. Therefore $\alpha \circ \beta \in M$ and, by lemma 5.16, α and β commute. ∎

6 Connectives in M-algebras

6.1 Connectives for arbitrary measurements

At this point two remarks are in order.

First, what we called connectives so far are actually algebraic operations on the set of measurements of an M-algebra. We will make the connection between these operations and proper connectives, i.e. the connectives of a propositional language in section 7.

Second, in our general definition of an M-algebras we have only one connective, namely negation. This is in contrast to the Birkhoff-von Neumann approach where we start with Hilbert space and with all classical propositional connectives given by the lattice operations of the lattice of projections. One reason for this is a methodological one. We can study important properties of M-algebras without considering other connectives. Another reason is that negation defined for arbitrary measurements shares important properties with classical negation and seems to be a good connective to start with. In the next chapter we will, step by step, consider the other propositional connectives defined between arbitrary measurements.

It is worth pointing out that for arbitrary M-algebras we *can* define all connectives in a reasonable way if we restrict them to commuting measurements. It turns out that the connectives thus defined behave classically. We prove for general M-algebras a theorem which may be regarded as a generalisation of the fact that commuting projections in Hilbert space form a Boolean algebra.

6.2 Connectives for commuting measurements

We now define, step by step, all propositional connectives for commuting measurements for arbitrary M-algebras.

Negation

Negation asserts the existence of a negation for every measurement. Let us study the commutation properties of $\neg \alpha$.

LEMMA 5.18. $\forall \alpha, \beta \in M$, *if α commutes with β, then $\neg \alpha$ commutes with β.*

Proof. Assume α commutes with β. We will see that β preserves $\neg \alpha$. Let $x \in FP(\neg \alpha)$. We have $x \in Z(\alpha)$. But $(\alpha \circ \beta)(x) = (\beta \circ \alpha)(x)$. Therefore $0 = \alpha(\beta(x))$, $\beta(x) \in Z(\alpha)$ and $\beta(x) \in FP(\neg \alpha)$. We have shown that β preserves $\neg \alpha$. By **Composition**, $(\neg \alpha) \circ \beta \in M$ and, by lemma 5.16, $\neg \alpha$ commutes with β. ∎

COROLLARY 5.19. $\forall \alpha, \beta \in M$, *$\alpha$ and β commute iff $\neg \alpha$ and β commute iff α and $\neg \beta$ commute iff $\neg \alpha$ and $\neg \beta$ commute.*

6. CONNECTIVES IN M-ALGEBRAS

Proof. By lemma 5.18 and Corollary 5.7. ∎

Conjunction

We will now define a conjunction between commuting measurements.

DEFINITION 5.20. *For any commuting measurements $\alpha, \beta \in M$, the conjunction $\alpha \wedge \beta$ is defined by:* $\alpha \wedge \beta = \alpha \circ \beta = \beta \circ \alpha$.

By lemma 5.16, the conjunction, as defined, is indeed a measurement.

LEMMA 5.21. *For any commuting $\alpha, \beta \in M$, the conjunction $\alpha \wedge \beta$ is the unique measurement γ such that $FP(\gamma) = FP(\alpha) \cap FP(\beta)$.*

Proof. By lemmas 5.6 and 5.12. ∎

One immediately sees that conjunction among commuting measurements is associative, commutative and that $\alpha \wedge \alpha = \alpha$ for any $\alpha \in M$.

Let us now study the commutation properties of conjunction.

LEMMA 5.22. $\forall \alpha, \beta, \gamma \in M$, *that commute in pairs, $\alpha \wedge \beta$ commutes with γ.*

Proof.

$$(\alpha \wedge \beta) \circ \gamma = (\alpha \circ \beta) \circ \gamma = \alpha \circ (\beta \circ \gamma) = \alpha \circ (\gamma \circ \beta) =$$

$$(\alpha \circ \gamma) \circ \beta = (\gamma \circ \alpha) \circ \beta = \gamma \circ (\alpha \circ \beta) = \gamma \circ (\alpha \wedge \beta)$$

∎

Disjunction

One may now define a disjunction between two commuting measurements in the usual, classical, way.

DEFINITION 5.23. *For any commuting measurements $\alpha, \beta \in M$, the disjunction $\alpha \vee \beta$ is defined by:* $\alpha \vee \beta = \neg(\neg \alpha \wedge \neg \beta)$.

By Corollary 5.19, the measurements $\neg \alpha$ and $\neg \beta$ commute, therefore their conjunction is well-defined and the definition of disjunction is well-formed.

The commutation properties of disjunction are easily studied.

LEMMA 5.24. $\forall \alpha, \beta, \gamma \in M$ *that commute in pairs, $\alpha \vee \beta$ commutes with γ.*

Proof. Obvious from Definition 5.23 and lemmas 5.18 and 5.22. ∎

The following is easily proved: use Definition 5.23, **Negation** and lemmas 5.10, 5.6 and 5.16.

LEMMA 5.25. *For any commuting measurements, α and β, their disjunction $\alpha \vee \beta$ is the unique measurement γ such that $Z(\gamma) = Z(\alpha) \cap Z(\beta)$.*

LEMMA 5.26. *If $\alpha, \beta \in M$ commute, then $FP(\alpha) \cup FP(\beta) \subset FP(\alpha \vee \beta)$.*

The inclusion is, in general, strict.

Proof. Since $Z(\alpha \vee \beta) \subset Z(\alpha)$, by lemma 5.10. ∎

Implication

Implication ($\rightarrow$) is probably the most interesting connective. It will play a central role in our treatment of connectives.

DEFINITION 5.27. For any commuting measurements $\alpha, \beta \in M$, the implication $\alpha \rightarrow \beta$ is defined by: $\alpha \rightarrow \beta = \neg(\alpha \wedge \neg \beta)$.

By Corollary 5.19, the measurements α and $\neg \beta$ commute, therefore their conjunction is well-defined and the definition of implication is well-formed.

The commutation properties of implication are easily studied.

LEMMA 5.28. *$\forall \alpha, \beta, \gamma \in M$ that commute in pairs, $\alpha \rightarrow \beta$ commutes with γ.*

Proof. Obvious from Definition 5.27 and lemmas 5.18 and 5.22. ∎

The following is easily proved: use Definition 5.27, **Negation** and lemmas 5.10, 5.6 and 5.16.

LEMMA 5.29. *For any commuting measurements, α and β, their implication $\alpha \rightarrow \beta$ is the unique measurement γ such that $Z(\gamma) = FP(\alpha) \cap Z(\beta)$.*

Lemma 5.29 characterises the zeros of $\alpha \rightarrow \beta$. Our next result characterises the fixed points of $\alpha \rightarrow \beta$ in a most telling and useful way.

LEMMA 5.30. *For any commuting measurements, α and β, their implication $\alpha \rightarrow \beta$ is the unique measurement γ such that $FP(\gamma) = \{x \in X \mid \alpha(x) \in FP(\beta)\}$.*

Proof. Assume α and β commute, and $x \in X$. Now, $\alpha(x) \in FP(\beta)$ iff (by **Negation**) $\alpha(x) \in Z(\neg \beta)$ iff $(\alpha \circ (\neg \beta))(x) = 0$ iff (by Definition 5.20 and lemma 5.18) $(\alpha \wedge (\neg \beta))(x) = 0$ iff $x \in Z(\alpha \wedge (\neg \beta))$ iff (by **Negation**) $x \in FP(\neg(\alpha \wedge (\neg \beta)))$ iff (by Definition 5.27) $x \in FP(\alpha \rightarrow \beta)$. ∎

The following is immediate.

COROLLARY 5.31. *For any commuting measurements α and β, if $x \in FP(\alpha)$ and $x \in FP(\alpha \rightarrow \beta)$, then $x \in FP(\beta)$.*

One may now ask whether the propositional connectives we have defined amongst commuting measurements behave classically. In particular, assuming that measurements α, β and γ commute in pairs, does the distribution law hold, i.e., is it true that $(\alpha \vee \beta) \wedge \gamma = (\alpha \wedge \gamma) \vee (\beta \wedge \gamma)$.

7 Amongst commuting measurements connectives are classical

It is now time to make the connection between the connectives as algebraic operations on M and the connectives of propositional logic. Let $\mathcal{L}$ be a language of propositional logic as defined in chapter 1 with the connective $\neg$, $\wedge$ $\vee$ and $\rightarrow$ being defined as usual. Let $\langle X, M \rangle$ be an M-algebra and $A \subset M$ be a set of pairwise commuting measurements. Let $\Psi : \mathcal{L} \rightarrow A$ be a function such that $\Psi(\neg \alpha) = \neg \Psi(\alpha)$, $\Psi(\alpha \rightarrow \beta) = \Psi(\alpha) \rightarrow \Psi(\beta)$. Note that the connectives on the left hand side are those of the language $\mathcal{L}$ and those of the right hand side are those of M-algebras.

The reader should not be confused by the fact that we use the same symbols for the connectives proper and the corresponding algebraic operations.

Our goal is the following

THEOREM 5.32. *Let $\langle X, M \rangle$ be an M-algebra and $A \subset M$ be a set of pairwise commuting measurements. Let Ψ be as above. If α is a classical propositional tautology, then $FP(\Psi(\alpha)) = X$.*

For the proof we use our system of classical propositional logic of chapter 1.

We will show that conjunction and disjunction (in M-algebras) may be defined in terms of negation and implication as usual. The proof will then proceed in seven steps: Modus Ponens, the four axiom schemes, conjunction and disjunction.

LEMMA 5.33. *For any commuting measurements α and β, if $FP(\alpha) = X$ and $FP(\alpha \rightarrow \beta) = X$, then $FP(\beta) = X$.*

Proof. By Corollary 5.31. ∎

LEMMA 5.34. *For any commuting measurements α and β,*

$$FP(\alpha \rightarrow (\beta \rightarrow \alpha)) = X.$$

Proof. Since α and β commute, for any $x \in X$: $\beta(\alpha(x)) = \alpha(\beta(x))$, therefore, by **Idempotence**, we have $\beta(\alpha(x)) \in FP(\alpha)$. By lemma 5.30, for any x, $\alpha(x) \in FP(\beta \rightarrow \alpha)$. By the same lemma: $x \in FP(\alpha \rightarrow (\beta \rightarrow \alpha))$. ∎

LEMMA 5.35. *For any pairwise commuting measurements α, β and γ*

$$FP((\alpha \rightarrow (\beta \rightarrow \gamma)) \rightarrow ((\alpha \rightarrow \beta) \rightarrow (\alpha \rightarrow \gamma))) = X.$$

Proof. By lemma 5.30, it is enough to show that for any $x \in X$, if $y = (\alpha \to (\beta \to \gamma))(x)$, then, if we define $z = (\alpha \to \beta)(y)$, and define $w = \alpha(z)$, we have: $\gamma(w) = w$. But since all the measurements above commute, by **Idempotence**, the state w satisfies $\alpha \to (\beta \to \gamma)$, $\alpha \to \beta$ and α. By Corollary 5.31, w satisfies β and $\beta \to \gamma$. For the same reason w satisfies γ. ■

LEMMA 5.36. *For any commuting measurements α and β, $FP(\alpha \to (\neg \alpha \to \beta)) = X$*

Proof. Given any state x. We need to show that $y = \neg \alpha(\alpha x) \in FP(\beta)$. But this obvious since $y = 0$. ■

LEMMA 5.37. *For any commuting measurements α and β we have $\alpha \to \beta = \neg \beta \to \neg \alpha$.*

Proof. By lemma 5.29 we have $Z(\alpha \to \beta) = Z(\beta) \cap FP(\alpha)$. In order to see the claim consider that $Z(\beta) = FP(\neg \beta)$ and $FP(\alpha) = Z(\neg \alpha)$ and apply lemma 5.29 again. ■

LEMMA 5.38. *For any commuting measurements α and β,*

$$FP((\alpha \to \beta) \to ((\neg \alpha \to \beta) \to \beta)) = X.$$

Proof. By lemma 5.30, it is enough to show that for any $x \in X$, if $y = (\alpha \to \beta)(x)$, then, if we define $z = (\neg \alpha \to \beta)(y)$ then we have: $\beta(z) = z$. But since all the measurements above commute, by **Idempotence**, the state z satisfies $\alpha \to \beta$ and $\neg \alpha \to \beta$. Then by lemma 5.37 z satisfies $\neg \beta \to \alpha$ and $\neg \beta \to \neg \alpha$. It follows that $\neg \beta(z) \in FP(\alpha)$ and $\neg \beta(z) \in FP(\neg \alpha)$. Therefore $\neg \beta(z) = 0$. It follows that $\beta(z) = z$. ■

LEMMA 5.39. *For any commuting measurements α and β, $\alpha \wedge \beta = \neg(\alpha \to \neg \beta)$.*

Proof.

$$FP(\neg(\alpha \to \neg \beta)) = Z(\alpha \to \neg \beta) = FP(\alpha) \cap Z(\neg \beta) = FP(\alpha) \cap FP(\beta).$$

By **Negation**, lemma 5.29 and **Negation**. The conclusion then follows from lemma 5.21. ■

LEMMA 5.40. *For any commuting measurements α and β, $\alpha \vee \beta = (\neg \alpha) \to \beta$.*

8. SEPARABLE M-ALGEBRAS

Proof.
$$Z((\neg\alpha) \to \beta) = FP(\neg\alpha) \cap Z(\beta) = Z(\alpha) \cap Z(\beta).$$

By lemma 5.29 and **Negation**. The conclusion then follows from Lemma 5.25. ∎

Considering the axiom system of chapter 1 we can now put the proof of Theorem 5.32 together from lemmas 5.33, 5.34, 5.35, 5.36 and 5.38. Note that for the application of lemma 5.34 we have $\Psi(\alpha \to (\beta \to \alpha)) = \Psi(\alpha) \to (\Psi(\beta) \to \Psi(\alpha))$ and the corresponding identities for the other lemmata.

8 Separable M-algebras

LEMMA 5.41. *In a separable M-algebra, a measurement is classical if and only if it commutes with any measurement.*

Proof. Suppose α is classical. Consider any $x \in X$ and any $\beta \in M$. Since α is classical we know that $x \in FP(\alpha)$ or $x \in Z(\alpha)$ and $\beta(x) \in FP(\alpha)$ or $\beta(x) \in Z(\alpha)$. If $x \in FP(\alpha)$, by Lemma 5.8, $\beta(x) \in Z(\alpha)$ implies $x \in Z(\beta)$ and $(\alpha \circ \beta)(x) = 0 = (\beta \circ \alpha)(x)$. But $\beta(x) \in FP(\alpha)$ implies $(\alpha \circ \beta)(x) = \beta(x) = (\beta \circ \alpha)(x)$.

If $x \in Z(\alpha)$ and $\beta(x) \in FP(\alpha)$, by Lemma 5.9, $\beta(x) = 0$ and $(\alpha \circ \beta)(x) = 0 = (\beta \circ \alpha)(x)$. If $\beta(x) \in Z(\alpha)$, then $(\alpha \circ \beta)(x) = 0 = (\beta \circ \alpha)(x)$.

Suppose, now that α commutes with any measurement β. By contradiction, assume $\alpha(x) \neq 0$ and $\alpha(x) \neq x$. By **Separability** there is some measurement γ such that $x \in FP(\gamma)$ and $\alpha(x) \notin FP(\gamma)$. But α and γ commute and: $(\alpha \circ \gamma)(x) = (\gamma \circ \alpha)(x) = \alpha(x)$. We see that $\alpha(x) \in FP(\gamma)$, a contradiction. ∎

Note that a measurement α is classical (see Definition 5.4) iff $Def(\alpha) = X$.

LEMMA 5.42. *If α is classical, so is $\neg\alpha$. If α and β are classical, then so are $\alpha \wedge \beta$, $\alpha \vee \beta$ and $\alpha \to \beta$.*

Proof. If α is classical, $Def(\alpha) = X$ and therefore $Def(\neg\alpha) = X$. For conjunction $(\alpha \wedge \beta)(x) = (\alpha \circ \beta)(x) = (\beta \circ \alpha)(x)$. If either $\alpha(x)$ or $\beta(x)$ is 0 then $(\alpha \wedge \beta)(x) = 0$, otherwise $\alpha(x) = x = \beta(x)$ and $(\alpha \wedge \beta)(x) = x$. The definitions of disjunction and implication in terms of negation and conjunction, then ensure the claim. ∎

CHAPTER 6

THE LOCAL VIEWPOINT: STATES AS LOGICAL ENTITIES

1 What can logic do about quantum mechanics?

The question whether we need a 'new logic' in order to reason properly in quantum theory is asked frequently. Do we have to depart from classical logic in building 'quantum logic' and if so, how? The answer that most physicists give to this question is that we do not. In fact, physicists put quantum mechanics to good use in an unprecedentedly successful way, and in this do they not use classical logic? In [49] Popper denies any need to depart from classical logic in order to reason properly in quantum mechanics. In view of recent developments in quantum logic as well as in the rapidly growing field of quantum computation it seems to us, however, that there are reasons to believe that the Hilbert space formalism of quantum mechanics encodes resources of a new kind which are of interest from the point of view of deduction and computation. However, in this book we pursue different logical aspects of the Hilbert space formalism and we are indifferent with respect to the above question.

Why did the question arise at all? As we already saw, the question of 'the logic of quantum mechanics' was, in the scientific literature, first raised by Birkhoff and von Neumann in their seminal 1936 paper. As discussed at length in chapter 4, their motivation for trying to discover the 'logic of quantum mechanics' was the fact that they considered the novel features of quantum mechanics such as the uncertainty relations to be logical in nature. Since these features are not reflected in classical logic, there is, according to Birkhoff-von Neumann, a need to construct a (logical) 'calculus' in which they are actually represented.

Later on it was Putnam, Finkelstein and others who put forward a view of quantum logic which for some time attracted considerable attention. Central to this paradigm is the idea that logic may be empirical. Putnam and his followers argued that the role of logic in quantum mechanics was similar to that of geometry in the theory of relativity. In the theory of relativity Euclidean geometry, which in Newtonian physics was still considered a priori, had to be revised on empirical grounds. In quantum mechanics, Putnam argued, it is (classical) logic that needs revision on empirical grounds. Similar to the way the theory of relativity teaches us the 'real' geometry quantum mechanics teaches us the 'real' logic. This is undoubtedly an attractive idea which, however, we do not pursue in this book.

In this book we adopt a different attitude, which is already implicit in the Birkhoff-von Neumann paper. Let us recall what they write in the Introduction: "The object of the present paper is to discover what logical structure one may hope to find in physical theories which, like quantum mechanics, do not conform to classical logic".

In fact, this is the task we pose ourselves: searching for logical structures in quantum mechanics. The procedure is this. We take a close look at Hilbert space and, as a result, identify and study certain logical structures implicit in Hilbert space. We then pursue the question whether these logical structures represent essential features of the formalism of quantum mechanics.

Can the structures we found shed light on the formalism of quantum mechanics? We think that they in fact can. So the answer to the question asked in the title of this section is that logic (as a science) can detect and study logical structures in the formalism of quantum mechanics which are relevant to the understanding of the formalism itself.

Are there any guidelines that may help us in our search for these structures? Are there any traits of quantum mechanics itself that could suggest certain directions of investigation? Let us speculate a bit about this.

As a good starting point we may look at the relationship between classical and quantum mechanics. We may start by analysing the way how quantum mechanics departs from classical mechanics. Given that quantum mechanics, as is often claimed more or less vaguely, does not conform to classical logic, then it is reasonable to ask how the transition from classical to quantum mechanics is reflected in the logical structures we are looking for.

There are various ways of viewing the relationship between classical and quantum mechanics. Since in classical mechanics we have no uncertainty relations it is the uncertainty relations that are often regarded as constituting the essential difference. Another crucial difference concerns the role of measurement. In classical mechanics a measurement does not involve a change of the state of the system measured. The fact that in quantum mechanics measurement does, in general, involve such a change of state is undoubtedly an essential difference between classical and quantum mechanics. Classical mechanics is often considered to be a limiting case of quantum mechanics as Newtonian mechanics is a limiting case of the theory of relativity. We may ask the question how these observations are reflected in the logical

1. WHAT CAN LOGIC DO ABOUT QUANTUM MECHANICS?

structures we find. What are uncertainty relations from the logical point of view? We already remarked in chapter 1 that, logically, the presence of uncertainty relations is reflected as nonmonotonicity of the logical structures implicit in the formalism of quantum mechanics. In chapter 5 we took the fact that there may be a change of state in quantum measurement as an inspiration for the dynamic view of propositions as acting on states rather than just being true or false in them.

There is, however, a general feeling expressed in a vast body of literature, popular scientific and seriously scientific or philosophical alike, that the fairly obvious differences between classical and quantum mechanics mentioned above are not the whole story. Rather the general impression seems to be that the way how quantum mechanics departs from classical mechanics touches on deeper ground. In the next chapter we will discuss the famous Einstein-Podolsky-Rosen (EPR) argument put forward in their famous paper entitled "Can the quantum-mechanical description of reality be considered complete?" [15]. In the EPR argument the term 'element of reality' plays a crucial role. EPR take it for granted that (physical) reality is to be viewed as consisting of separate 'elements of reality'. And, in fact, once this fragmenting view of reality is accepted, it is hard to avoid the EPR conclusion that quantum mechanics does not provide a complete description of physical reality. Therefore, if quantum mechanics is in fact a complete description of physical reality as seems to be generally assumed nowadays, then something must be wrong with this view of reality. It seems that the way quantum mechanics departs from classical mechanics is of an even more profound nature than the way the (special) theory of relativity departs from classical mechanics. In the latter case we 'just' have to abandon our views on space and time. In the case of quantum mechanics it seems that we have to abandon our views on the very nature of reality. This is all pervading the literature on the foundations of quantum mechanics be it popular scientific or seriously philosophical. It is the intuition of oneness, interconnectedness and wholeness, which is prevalent in Eastern thought for instance, that finds strong support in quantum mechanics. But this is the realm of intuition and metaphor, perhaps even of philosophy, and it is hard to make something scientific of this at the level of ordinary discourse.

How can we, at the level of logic, reflect the shift in our perception of reality which is forced upon us in the transition from classical mechanics to quantum mechanics? A possible answer is this. Classical mechanics and classical logic conform to each other and the view of reality that underlies classical mechanics also underlies classical logic. If, as seems to be the case, our 'classical' view of reality is to be revised in quantum mechanics, we must ask the question whether logic, i.e. the quantum logic to be constructed, can account for quantum mechanics if it does not reflect this shift. We will in chapter 8 describe a way of departing from classical logic for the sake of quantum logic which may be regarded as reflecting this intuition.

2 States as logical entities

The concept of a state of a physical system plays a role both in classical and in quantum mechanics. In our study of M-algebras, which constitute our first abstraction from the Hilbertian formalism, we treated the concept of a state of a physical system as a primitive notion. In this chapter we ask the question what *is* a state from the logical point of view. Can we view the states of a physical system as logical entities themselves and if so, what is the nature of these logical entities?

In classical physics, the concept of a state is, from the logical point of view, unproblematic. Logically, in classical physics a state is a complete classical theory. It can be identified with the set of all physical statements *true* about the system. In this sense the state of the system at a certain point in time fully contains all the information about the system. What in classical mechanics is particularly convenient is the fact that once we know the momenta and the positions of the particles constituting the system, we know all relevant physical properties. Therefore, from the logical point of view, a state can be described by a single proposition, namely by the proposition specifying all values of the momenta and positions at a given time. From this we can then compute (deduce) the values of all relevant physical quantities. This is what in classical mechanics is known as *phase space*. So, the logical analogue of the concept of a state in classical mechanics is that of a *complete classical theory*.

This simple concept of a state is based on the view underlying classical mechanics that a physical system *possesses* certain properties and does *not possess* others. The propositions expressing the physical properties a physical system can possess according to classical mechanics may be viewed as having the form $A = \mu$, where A denotes a physical quantity (observable) such as position, momentum, energy etc. In any given state, for any observable A the proposition $A = \mu$ is true for exactly one (real) value μ. For any value $\rho \neq \mu$ the proposition $A = \rho$ is false which is equivalent to $\neg(A = \rho)$. For any physically meaningful property the system either possesses it or not. Any given proposition holds or does not hold at any given point in time. In the latter case it is, by classical logic, the negation of the proposition that holds. Take for instance position Q and consider the proposition $Q = \mu$, take moreover momentum P and consider the proposition $P = \lambda$. Then according to classical mechanics these propositions or their negations are true. We may for instance have $Q = \mu$ true and $P = \lambda$ true. In this case

2. STATES AS LOGICAL ENTITIES

we can deduce any other proposition true about the system. We may for instance infer a proposition of the form $\neg(E = \rho)$, where E denotes kinetic energy. Obviously, this concept of a state rests on classical logic and in particular on the notion of truth underlying classical logic.

How can we, in classical physics, know if the system possesses a certain property, how can we know that, say $A = \mu$ is true? The answer is that given this proposition we can at least in principle find out whether it is true via *measurement*. $A = \mu$ is true if and only if a measurement of the physical quantity A yields the value μ. Whenever we measure A, we get μ and no other value as a result of measurement.

Now, what's different in quantum mechanics? Why can't we represent the state of a quantum system analogously, namely by the set of those propositions that are true or false in this state? The reason is that, in quantum mechanics, the term 'is true' is far less clear. In classical mechanics we said that 'to be true' may, roughly, be taken as 'being measured' or at least 'to be measurable'. In quantum mechanics things aren't that simple. Given a state x in quantum mechanics and a proposition $A = \mu$. Suppose we perform a measurement of an observable A in x. Then the following three cases may occur. First, the probability to measure μ is 1 in which case we may reasonably say that $A = \mu$ is true (in state x). Second, the probability to get μ as a result of measurement may be 0. In this case we may reasonably say that $A = \mu$ is false or, equivalently, $\neg(A = \mu)$ is true. In quantum mechanics there is, however, a third case which marks the difference with classical mechanics. Namely, the probability to get μ may be greater than zero and smaller than one. Let us for the moment call these propositions contingent with respect to x. It is then obviously insufficient to represent the state x by the set of those propositions that are true or false in x because this does not give us any information about the contingent propositions and their probabilities. It seems that a proper representation of a quantum state must specify probabilities. In a purely logical treatment of the concept of a physical state we should, however, try to avoid specifying probabilities.

Let us now reflect on the problem of representing states within the framework of M-algebras. Given an M-algebra $\langle X, M \rangle$ and a state $x \in X$. Again, it is insufficient to represent x as the set of those propositions α such that $x \in FP(\alpha)$ or $x \in FP(\neg\alpha)$. In fact, there exist, generally, propositions such that neither $x \in FP(\alpha)$ nor $x \in FP(\neg\alpha)$. These propositions act on x in that they neither leave it unchanged nor send it to zero. They are neither 'true' nor 'false'.

We may think of the action of propositions on states in an M-algebra as a sort of *coming true* rather than *being true*. We may say that α *is* true in x if $x \in FP(\alpha)$ and x is false in x if $x \in FP(\neg\alpha)$. Otherwise, i.e. in case that $\alpha(x) = y \neq x$ and $\alpha(x) \neq 0$ we may say that α *comes* true in x. Thus α *comes* true in x if it *is* true in $\alpha(x)$. Hence the representation of the state x must give us information not just on what is true or false in x but about what comes true in x. Thus in an M-algebra it is the coming true of a proposition that replaces or generalises the being true of a proposition in classical logic. This is, in an M-algebra, the dynamic analogue of the static concept of being true in classical logic. However, coming true in x involves a different state which in turn must be specified. Thus, intuitively, we must require the logical entity representing a state x as also specifying other states, namely all those states in which a proposition *is* true when it *comes* true in state x.

Technically speaking, it is as follows. The logical entity representing a (physical) state x in classical logic, namely a complete theory, contains all propositions true or false in state x. When, however, we are concerned with propositions that act on the state x or, as we said, have the property of coming true rather than being true in state x, then the logical representation of x must encode all propositions that come true at x. In other words, the logical representation of a state x must encode the action of the propositions on x. The action of a proposition on x, however, yields a new state y, and therefore the state x must encode other states. So we inevitably hit here on the phenomenon of *encodedness of states in other states* which will play a dominant role in our study of *holistic logics* introduced in chapter 8. We will see that, there, a state is a logical entity that encodes *other states* and also itself.

3 Implication M-algebras

In our study of M-algebras we dealt for the most part with what we call *global properties*. This also applies to the Birkhoff-von Neumann paper. By a global property we mean a property that does not depend on some fixed state. In chapter 5 we proved that in any M-algebra commuting propositions obey classical logic. This is a global property.

In this section we focus on what we call *local properties*. This is what we mean by *local viewpoint*. By a local property we mean a property concerned with a certain fixed state. Some of the properties we already dealt with in connection with M-algebras *are* local properties. Recall for instance the axiom of Interference. The study of local properties of M-algebras gives us insight into nature of the states themselves. We regard the local viewpoint as constituting the main difference between the approach to quantum logic presented in this book and Birkhoff-von Neumann quantum logic.

The concept of an M-algebra as introduced and studied in chapter 5 is

3. IMPLICATION M-ALGEBRAS

still too general for an investigation of this sort. For this we need to consider more special structures. We will now introduce M-algebras which we call *Implication M-algebras*.

In introducing Implication M-algebras and later on *Conjunction M-algebras* we have to keep the following in mind. As we saw in the last chapter we *can*, in any M-algebra, introduce all the propositional connectives in a natural way if we restrict ourselves to commuting measurements. We therefore have to keep in mind that whenever we introduce the connectives for arbitrary measurements we have to do this in such a way that for commuting measurements they agree with the connectives already defined.

In an Implication M-algebra we allow for another connective defined between arbitrary measurements, namely implication, which we denote by $\leadsto$. This connective has the following intuitive meaning: $\alpha \leadsto \beta$ says: "If we measure α, we (also) get β".

In chapter 5 we proved the following.

Let $\langle X, M \rangle$ be an M-algebra, $x \in X$ and $\alpha, \beta \in M$ be two commuting measurements. Then we have

$$x \in FP(\alpha \to \beta) \text{ iff } \alpha(x) \in FP(\beta).$$

Given an M-algebra $\langle X, M \rangle$ and consider a fixed state $x \in X$. We now define a binary relation $\vdash_x$ between arbitrary measurements as follows.

DEFINITION 6.1. Let $\langle X, M \rangle$ be an M-algebra and $x \in X$. Given $\alpha, \beta \in M$. Then we say $\alpha \vdash_x \beta$ iff $\alpha(x) \in FP(\beta)$.

We denote the set of these relations by $\mathcal{C}$, i.e. $\mathcal{C} =: \{\vdash_x | \ x \in X\}$. We may, intuitively, think of these relations as consequence relations although in the general situation of M-algebras we cannot expect them to satisfy the minimal conditions of chapter 1. But we will see in chapter 9 that in the case of a Hilbert space these conditions are in fact satisfied and thus these binary relations may rightfully be called consequence relations.

In the sequel we write $\vdash_x \alpha$ for $\top \vdash_x \alpha$.

LEMMA 6.2. $\alpha \vdash_x \beta$ iff $\vdash_{\alpha(x)} \beta$ iff $\alpha(x) \in FP(\beta)$

Proof. By definition. ∎

DEFINITION 6.3. We call an M-algebra $\langle X, M \rangle$ an Implication M-algebra if there exists a function $I : M \times M \to M$ such that for any $x \in X$, $\alpha \vdash_x \beta$ iff $x \in FP(I(\alpha, \beta))$ or equivalently $\alpha(x) \in FP(\beta)$.

Note that the function I is, if it exists, unique because a measurement is uniquely determined by its set of fixed points. Call it the implication of the M-algebra. We also write $\alpha \leadsto \beta$ for $I(\alpha, \beta)$. We have by definition that $\alpha \mathrel{\vert\kern-0.3em\sim}_x \beta$ iff $\alpha \rightsquigarrow \beta$. In the terminology introduced later on we defined implication to be an internalising connective for all $\mathrel{\vert\kern-0.3em\sim} \in \mathcal{C}$.

Recall that any orthomodular space and in particular any Hilbert space gives rise to an Implication M-algebra in a natural way. By Proposition 2.26 the implication I is given by $I(\alpha, \beta) = \alpha^\perp \vee (\alpha \wedge \beta)$, i.e. the . This is not accidental, as we will see. For commuting measurements the implication agrees with that already defined.

LEMMA 6.4. *If α and β commute we have $\alpha \rightsquigarrow \beta = \alpha \rightarrow \beta$.*

Proof. By the definition of $\rightsquigarrow$ and Lemma 5.30 ∎

4 Conjunction M-algebras

DEFINITION 6.5. We call an M-algebra $\langle X, M \rangle$ a Conjunction M-algebra if there exists a function $C : M \times M \rightarrow M$ such that for any $\alpha, \beta \in M$ we have $FP(C(\alpha, \beta)) = FP(\alpha) \cap FP(\beta)$. We call such a function a conjunction. For $C(\alpha, \beta)$ we also write $\alpha \wedge \beta$.

Again, if an M-algebra admits a conjunction, this conjunction is unique because measurements are uniquely determined by their set of fixed points.

The following lemma follows from the definition of a conjunction and lemma 5.21.

LEMMA 6.6. *Let $\mathcal{A}$ be an M-algebra and C a conjunction of $\mathcal{A}$. Then for commuting measurements C agrees with the conjunction (already) defined.*

THEOREM 6.7. *Given a Conjunction M-algebra $\mathcal{A} = \langle X, M \rangle$ with conjunction C. Then $\mathcal{A}$ admits a (unique) implication I. Namely implication is the Sasaki hook: $I(\alpha, \beta) = \neg \alpha \vee (\alpha \wedge \beta) = \alpha \rightarrow (\alpha \wedge \beta)$.*

Note that for any $\alpha, \beta \in M$ $\neg \alpha \vee (\alpha \wedge \beta)$ is defined. Namely we have $FP(\alpha) \subset FP(\alpha \wedge \beta)$. Hence α and $\alpha \wedge \beta$ commute by lemma 5.17 and thus by lemma 5.18 $\neg \alpha$ and $\alpha \wedge \beta$ commute. The theorem says that any Conjunction M-algebra is also an Implication M-algebra.

Proof. It suffices to prove that $FP(\alpha \rightarrow \beta) = \{x \mid \alpha(x) \in FP(\beta)\}$. By Lemma 5.30 we have $FP(\alpha \rightarrow \beta) = \{x \mid \alpha(x) \in FP(\alpha \wedge \beta)\}$. By the definition of conjunction this set is equal to $\{x \mid \alpha(x) \in FP(\alpha) \cap FP(\beta)\}$. Since $\alpha(x) \in FP(\alpha)$, it follows that $FP(\alpha \rightarrow \beta) = \{x \mid \alpha(x) \in FP(\beta)\}$. ∎

Recall from chapter 2 that the connective $\rightsquigarrow$ defined by $\alpha \rightsquigarrow \beta = \neg \alpha \vee (\alpha \wedge \beta)$ is called the *Sasaki hook* also denoted by $\rightsquigarrow_s$. Note that the Sasaki hook is classically equivalent to material implication, i.e, $\vdash (\alpha \rightsquigarrow \beta) \leftrightarrow (\alpha \rightarrow \beta)$.

5 Strongly separable M-algebras

5.1 Basic properties

As a matter of convention let us state here that whenever we fix an element x of an M-algebra we mean a non-zero element if not stated otherwise.

DEFINITION 6.8. Let $\langle X, M \rangle$ be an M-algebra and $x \in X$. We call a measurement e a pointer to x if $FP(e) = \{x, 0\}$

Again, note that given x then a pointer to x is unique. We denote it by e_x.

LEMMA 6.9. *Let $\langle X, M \rangle$ be an M-algebra, let e_x be the pointer to x. Then for any $x, y \in X$ we have $e_x(y) = 0$ for $y \in Z(e_x)$, otherwise $e_x(y) = x$.*

Proof. The first claim expresses a familiar property of negation. For the second claim suppose that not $y \in Z(e_x)$. Then $e_x(y) \neq 0$ and $e_x(y) \in FP(e_x)$. But since $FP(e_x) = \{x, 0\}$, it follows that $e_x(y) = x$. ∎

DEFINITION 6.10. We call an M-algebra $\langle X, M \rangle$ strongly separable if every $x \in X$ has a pointer.

PROPOSITION 6.11. *A strongly separable M-algebra $\langle X, M \rangle$ is separable.*

Proof. Given two distinct $x, y \in X$. Then $x \in FP(e_x)$ and not $y \in FP(e_y)$ and vice versa. It follows that $\langle X, M \rangle$ is separable. ∎

PROPOSITION 6.12. *In a strongly separable M-algebra the map $\varphi : X \to \mathcal{C}$ defined by $\varphi(x) =: \hspace{-0.1em}\sim_x$ is a bijection. Any measurement α induces a map $\bar{\alpha} : \mathcal{C} \to \mathcal{C}$ such that $\bar{\alpha}(\sim_x) = \sim_{\alpha(x)}$.*

Proof. Clearly, φ is a surjective mapping. In order to see that it is injective let x and y be two distinct states. Then we have $\sim_x \neq \sim_y$ because $\sim_x e_x$ and $\not\sim_y e_x$. Hence $\varphi(x) \neq \varphi(y)$. ∎

LEMMA 6.13. *Given a strongly separable M-algebra $\langle X, M \rangle$. Let $x, y \in X$. Then $e_x(y) = 0$ implies $e_y(x) = 0$*

Proof. The claim follows from **Interference**. Namely suppose $e_x(y) = 0$. Then we have $e_x(e_y(x)) = 0$ and thus $e_y(e_x(e_y(x))) = e_x(e_y(x))$. Note that $e_x = x$. By **Interference** we get that $e_y(x) \in FP(e_x)$. Hence $e_y(x) = 0$. ∎

DEFINITION 6.14. Let $\langle X, M \rangle$ be a strongly separable M-algebra. Let $x, y \in X$. Then we say that x and y are orthogonal if $e_x(y) = 0$.

By lemma 6.13 the relation of orthogonality is symmetric.

In view of the above the following proposition is obvious.

PROPOSITION 6.15. *The separable M-algebra of an orthomodular space is a strongly separable M-algebra and thus a strongly separable Implication M-algebra with the Sasaki hook as its implication.*

5.2 Encodedness

One result of our intuitive considerations at the beginning of this chapter was that quantum states viewed as logical entities should encode the action of propositions.

We have the

PROPOSITION 6.16. *Let $\langle X, M \rangle$ be a strongly separable Implication M-algebra, $x \in X$ (non-zero) and $\alpha \in M$. Then we have $\alpha(x) = y \neq 0$ iff $\alpha \mathrel{\mid\!\sim}_x e_y$ (or equivalently) $\mathrel{\mid\!\sim}_x \alpha \leadsto e_y$. We have $\alpha(x) = 0$ iff $\alpha \mathrel{\mid\!\sim}_x \bot$.*

Proof. Suppose $\alpha(x) = y$ with $y \neq 0$. It follows by the definition of a pointer that $\alpha(x) \in FP(e_y)$. By the definition of $\mathrel{\mid\!\sim}_x$ this says that $\alpha \mathrel{\mid\!\sim}_x e_y$. For the other direction let $\alpha \mathrel{\mid\!\sim}_x e_y$. This says that $\alpha(x) \in \{y, 0\}$ and, since $y \neq 0$, we have $\alpha(x) = y$.

Suppose $\alpha(x) = 0$. This is equivalent to $\mathrel{\mid\!\sim}_x \neg \alpha$, which in turn is equivalent to $\alpha \mathrel{\mid\!\sim}_x \bot$. ∎

LEMMA 6.17. *Let $\langle X, M \rangle$ be a strongly separable Implication M-algebra and $x \in X$. Then we have for any measurement α*

- *(i) $\mathrel{\mid\!\sim}_x \alpha$ iff $e_x \leadsto \alpha = \top$ and thus $\neg(e_x \leadsto \alpha) = \bot$*
- *(ii) $\mathrel{\mid\!\not\sim}_x \alpha$ iff $e_x \leadsto \alpha = \neg e_x$ and thus $\neg(e_x \leadsto \alpha) = e_x$*

Proof. For (i) suppose $\mathrel{\mid\!\sim}_x \alpha$. This says that $x \in FP(\alpha)$. We need to prove that for any $y \in X$ we have $e_x(y) \in FP(\alpha)$. But this is obvious because for any $y \in X$ we have either $e_x(y) = x$, in which case x and y are not orthogonal, or $e_x(y) = 0$ otherwise.

For the other direction, if $e_x \leadsto \alpha = \top$, we have that $x \in FP(e_x \leadsto \alpha$. This says that $x = e_x(x) \in FP(\alpha)$ or equivalently $\mathrel{\mid\!\sim}_x \alpha$.

For (ii) assume $\mathrel{\mid\!\not\sim}_x \alpha$. We need to show that $FP(e_x \leadsto \alpha) = Z(e_x)$. If $y \in Z(e_x)$, i.e. y is orthogonal to x, then $e_x \leadsto \alpha$ holds at y because $e_x(y) = 0$ and thus $e_x(y)) \in FP(\alpha)$. If y is non-orthogonal to x, i.e. y is not in $Z(e_x)$, we have $e_x(y) = x$. It follows by the hypothesis that $e_x(y) \notin FP(\alpha)$. This says that $e_x \leadsto \alpha$ does not hold at y. It follows that $FP(e_x \leadsto \alpha) = Z(e_x)$.

If, for the other direction, we have $e_x \leadsto \alpha = \neg e_x$, then we cannot have $\mathrel{\mid\!\sim}_x \alpha$ by (i). ∎

Another result of our intuitive considerations was that states should encode other states. This is expressed as follows.

THEOREM 6.18. *Let $\langle X, M \rangle$ be a strongly separable Implication M-algebra and let x, y be non-orthogonal states. Then we have*

- *(i)* $\alpha \mathrel{\mid\!\sim}_x \beta$ *iff* $e_x \mathrel{\mid\!\sim}_y \alpha \rightsquigarrow \beta$
- *(ii)* $\alpha \mathrel{\mid\!\not\sim}_x \beta$ *iff* $e_x \mathrel{\mid\!\sim}_y e_x \rightsquigarrow \neg(e_x \rightsquigarrow (\alpha \rightsquigarrow \beta))$

By symmetry the above holds if we interchange x and y.

Proof. Suppose x and y are non-orthogonal. Note that $e_x(y) = x$. (i) $\alpha \mathrel{\mid\!\sim}_x \beta$ is equivalent to $x \in FP(\alpha \rightsquigarrow \beta)$. Since $e_x(y) = x$, this equivalent to $e_x(y) \in FP(\alpha \rightsquigarrow \beta)$, which says $e_x \mathrel{\mid\!\sim}_x \alpha \rightsquigarrow \beta$.

For (ii) note that by 6.17 $\alpha \mathrel{\mid\!\not\sim}_x \beta$ is equivalent to $e_x \rightsquigarrow (\alpha \rightsquigarrow \beta) = \neg e_x$. The right hand side is then equivalent to $e_x \mathrel{\mid\!\sim}_y e_x \rightsquigarrow e_x$. But this is true because $e_x \rightsquigarrow e_x = \top$. ∎

The above says that non-orthogonal states encode each other. We will come across this phenomenon of mutual encodedness of states again in chapter 8.

PROPOSITION 6.19. *Let $\langle X, M \rangle$ be a strongly separable Implication M-algebra. We have*

- *(i)* $\mathrel{\mid\!\sim}_x \alpha$ *iff* $\mathrel{\mid\!\sim}_x e_x \rightsquigarrow \alpha$
- *(ii)* $\mathrel{\mid\!\not\sim}_x \alpha$ *iff* $\mathrel{\mid\!\sim}_x \neg(e_x \rightsquigarrow \alpha)$
- *(iii) If* $\mathrel{\mid\!\not\sim}_x \alpha$ *and* $\mathrel{\mid\!\not\sim}_x \neg\alpha$, *then* $\mathrel{\mid\!\sim} \neg(e_x \rightsquigarrow \alpha)$ *but* $\alpha \mathrel{\mid\!\not\sim}_x \neg(e_x \rightsquigarrow \alpha)$.

Proof. Consider that x is not orthogonal to itself. Then (i) and (ii) follow from 6.18. For (iii) we need to see that $\alpha \mathrel{\mid\!\not\sim}_x \neg(e_x \rightsquigarrow \alpha)$. This would say that $\alpha(x) \in FP(e_x)$, which, however, is not the case because by the hypothesis $\alpha(x)$ is distinct both from x and 0. ∎

We may view $e_x \rightsquigarrow \alpha$ and $\neg(e_x \rightsquigarrow \alpha)$ as expressing meta-statements. Namely, we may, if we think of $\mathrel{\mid\!\sim}_x$ as a consequence relation, view $e_x \rightsquigarrow \alpha$ as saying 'α is provable in $\mathrel{\mid\!\sim}_x$' and $\neg(e_x \rightsquigarrow \alpha)$ as saying 'α is not provable in $\mathrel{\mid\!\sim}_x$'. (ii) and iii together express *nonmonotonicity*, namely (ii) says "If α is not provable, then it is provable that it is not provable" and (iii) says " but not from α".

6 No windows theorem: first version

We now consider again a language of propositional logic $\mathcal{L}$ built up from a set of variables and the above connectives — now viewed as genuine propositional connectives — as described in chapter 1. Consider such a language and a surjective function $\Psi : \mathcal{L} \to M$ such that $\Psi(\neg\alpha) = \neg\Psi(\alpha)$, $\Psi(\alpha \wedge \beta) = \Psi(\alpha) \wedge \Psi(\beta)$. This 'valuation' function is fixed in the sequel.

Recall that for an $x \in X$ we defined a binary relation $\mathrel{\vert\!\sim}_x$. This is a relation between elements of M. We now define the corresponding binary relation between formulas, again denoted by $\mathrel{\vert\!\sim}_x$, as follows

$$\alpha \mathrel{\vert\!\sim}_x \beta \text{ iff } \Psi(\alpha) \mathrel{\vert\!\sim}_x \Psi(\beta).$$

We call $\mathrel{\vert\!\sim}_x$ *monotonic* if $\mathrel{\vert\!\sim}_x \beta$ implies $\alpha \mathrel{\vert\!\sim}_x \beta$ for any α, otherwise *nonmonotonic*.

Of course, on the left hand side $\mathrel{\vert\!\sim}_x$ is the relation between the formulas α and β to be defined. The relation $\mathrel{\vert\!\sim}_x$ on the right hand side is the relation between measurements already defined. In what follows we mean by $\mathrel{\vert\!\sim}_x$ the relation defined above. By $\mathrel{\vert\!\sim}_x$ we mean $\top \mathrel{\vert\!\sim}_x \alpha$.

We define Σ_g as $\{\alpha \in \mathcal{L} \mid (\forall x \in X) \mathrel{\vert\!\sim}_x \alpha\}$. Intuitively, this is the 'global theory' of X. Note that $\alpha \in \Sigma_g$ iff $\Psi(\alpha) = \top$.

DEFINITION 6.20. *Let $\mathcal{A} = \langle X, M \rangle$ be an M-algebra and given a state x. We call x classical if for every measurement α we have $\alpha(x) = x$ or $\alpha(x) = 0$.*

LEMMA 6.21. *Let $\mathcal{A} = \langle X, M \rangle$ be a strongly separable Conjunction M-algebra and $x \in X$. Suppose that $\Sigma_g \cup \{\sigma_x\}$ is consistent. Then we have for any α*

$$\Sigma_g \cup \{\sigma_x\} \vdash \alpha \text{ iff } \mathrel{\vert\!\sim}_x \alpha.$$

Proof. First note that a Conjunction M-algebra admits a unique implication $\rightsquigarrow$, namely $\alpha \rightsquigarrow \beta = \neg\alpha \vee (\alpha \wedge \beta)$. Further recall that $\alpha \mathrel{\vert\!\sim}_x \beta$ iff $\mathrel{\vert\!\sim}_x \alpha \rightsquigarrow \beta$.

Assume $\mathrel{\vert\!\sim}_x \alpha$. Then we have $e_x \rightsquigarrow \alpha = \top$ by lemma 6.17 and thus $\sigma_x \rightsquigarrow \alpha \in \Sigma_g$. Clearly, we then have $\Sigma_g \vdash \sigma_x \rightsquigarrow \alpha$. Since $\rightsquigarrow$ is classically equivalent to $\rightarrow$, i.e. material implication, we get that $\Sigma_g \vdash \sigma_x \rightarrow \alpha$. By the deduction theorem of classical logic (see 1.8) we have $\Sigma_g \cup \{\sigma_x\} \vdash \alpha$. Thus the direction from right to left is proved.

For the other direction suppose

$$(1)\ \Sigma_g \cup \{\sigma_x\} \vdash \alpha$$

and

6. NO WINDOWS THEOREM: FIRST VERSION

$$(2)\ \not\hspace{-2pt}\sim_x \alpha.$$

Then (2) implies by proposition 6.19

$$(3)\ \not\hspace{-2pt}\sim_x \neg(\sigma_x \rightsquigarrow \alpha)$$

By the direction already proved we get

$$(4)\ \Sigma_g \cup \{\sigma_x\} \vdash \neg(\sigma_x \rightsquigarrow \alpha)$$

Since $\rightsquigarrow$ is classically equivalent to $\rightarrow$, we get

$$(5)\ \Sigma_g \cup \{\sigma_x\} \vdash \neg(\sigma_x \rightarrow \alpha)$$

But now it follows from (1) and the deduction theorem of classical logic that

$$(6)\ \Sigma_g \vdash \sigma_x \rightarrow \alpha$$

Hence

$$(7)\ \Sigma_g \cup \{\sigma_x\} \vdash \sigma_x \rightarrow \alpha$$

But (5) and (7) say that $\Sigma_g \cup \{\sigma_x\}$ is inconsistent contrary to the hypothesis. ■

LEMMA 6.22. *Let the hypothesis be as in the above lemma. Suppose that x is not classical. Then $\Sigma_g \cup \{\sigma_x\}$ is inconsistent.*

Proof. Assume $\Sigma_g \cup \{\sigma_x\}$ is consistent. Since x is not classical, there exists an α such that neither $\mid\hspace{-2pt}\sim_x \alpha$ nor $\mid\hspace{-2pt}\sim_x \neg\alpha$. By 6.19 we then have

$$(1)\ \mid\hspace{-2pt}\sim_x \neg(\sigma_x \rightsquigarrow \alpha)$$

and

$$(2)\ \alpha \not\hspace{-2pt}\sim_x \neg(\sigma_x \rightsquigarrow \alpha)$$

By the above lemma we then have

$$(3)\ \Sigma_g \cup \{\sigma_x\} \vdash \neg(\sigma_x \rightarrow \alpha)$$

By the monotonicity of classical logic we have

$$(4)\ \Sigma_g \cup \{\sigma_x\} \vdash \alpha \rightarrow \neg(\sigma_x \rightarrow \alpha)$$

Considering that $\rightsquigarrow$ is classically equivalent to $\rightarrow$ we have

$$(5)\ \Sigma_g \cup \{\sigma_x \vdash \alpha \rightsquigarrow \neg(\sigma_x \rightsquigarrow \alpha)$$

and by 6.21

$$(6)\ \mid\!\sim_x \alpha \rightsquigarrow \neg(\sigma_x \rightsquigarrow \alpha)$$

This means

$$(7)\ \alpha \mid\!\sim_x \neg(\sigma_x \rightsquigarrow \alpha)$$

But this contradicts (2). It follows that $\Sigma_g \cup \{\sigma_x\}$ is inconsistent. ∎

DEFINITION 6.23. Let $\mathcal{A} = \langle X, M \rangle$ be a strongly separable M-algebra. Suppose we have a family of states $(x_i)_{i \in I}$ such that no state is orthogonal to all x_i's, equivalently if $\bigcap_i FP(\neg e_{x_i})$ is empty. Then we call $(x_i)_{i \in I}$ a basis of $\mathcal{A}$. We call $\mathcal{A}$ finite-dimensional if it has a finite basis.

DEFINITION 6.24. Let $\mathcal{A} = \langle X, M \rangle$ be a Conjunction M-algebra. Then we call a formula φ a KS-tautology for $\mathcal{A}$ if if is a (classical) tautology and $\Psi(\varphi) = \bot$.

Remark: It should be pointed out here that the above definition of a *KS*-tautology does not capture all aspects of the famous tautology presented by Kochen and Specker in [34]. In that tautology only compatible propositions are combined via the propositional connectives. In the above definition we allow for *any* tautology.

LEMMA 6.25. *Let $\mathcal{A}$ be finite dimensional Conjunction M-algebra. Let $x_i, i = 1, ..., n$ be a (finite) basis. Then we have $FP(\neg \bigwedge_i \neg \sigma_i) = X$.*

Proof. We need to see that $FP(\bigwedge_i \neg \sigma_i) = \bigcap_i FP(\neg \sigma_i) = \{0\}$. Assume there is a non-zero state y in that intersection. This would mean that y is orthogonal to every $x_i, i = 1, .., n$ contrary to the assumption that $x_i, i = 1, ..., n$ is a basis. ∎

We have by the definition of a Conjunction M-algebra

LEMMA 6.26. *Σ_g is closed under conjunctions.*

We can now put the above lemmata together for the proof of the following theorem.

THEOREM 6.27. *Let $\mathcal{A} = \langle X, M \rangle$ be a finite-dimensional strongly separable Conjunction M-algebra without classical states. Then Σ_g is inconsistent. In fact there exists a KS-tautology for $\mathcal{A}$.*

COROLLARY 6.28. *Any finite dimensional orthomodular space and thus any finite-dimensional Hilbert space of dimension at least two admits a KS-tautology.*

Remark: We may view the above theorem as saying a bit more than just giving a sufficient condition for the existence of a KS-tautology in certain M-algebras. We may look at this theorem as follows. Given an Implication M-algebra $\mathcal{A}$ satisfying the hypotheses of the theorem. We would then like to define a conjunction in $\mathcal{A}$ such that implication becomes definable in terms of negation and conjunction as is the case in classical logic. The theorem then says that this cannot be done in a reasonable way in the sense that any 'classical' conjunction makes the global theory of $\mathcal{A}$ inconsistent. So we may view the theorem as saying that M-algebras of the above without classical states and on the one hand havin a consistent global theory and on the other hand admitting a 'classical' conjunction (and thus a 'classical' implication), do not exist.

Proof. Let $x_i, i = 1, ...n$ be a basis of $\mathcal{A}$. By lemma 6.22 we have

$$\Sigma_g \vdash \neg \sigma_{x_i} \text{ for } i = 1, ..n.$$

Thus

$$\Sigma_g \vdash \bigwedge_i \neg \sigma_{x_i}$$

On the other hand we have by lemma 6.25

$$\neg \bigwedge_i \neg \sigma_{x_i} \in \Sigma_g.$$

Therefore Σ_g is inconsistent. ∎

7 Limiting case theorem: first version

In this section we prove a limiting case theorem for M-algebras.

Classical mechanics is a limiting case of quantum mechanics. In which sense? In which sense do we have to 'pass to the limit from quantum mechanics' in order to get classical mechanics as the limit? There are various ways of intuitively viewing this process. One may for instance say that 'passing to the limit' means passing from the presence of uncertainty relations to the absence of uncertainty relations. Another way of looking at this is to say that this process is from change of state in measurement to the absence of change of state in measurement. Another version is to say that it is from non-commuting measurements to commuting measurements. In this section we study the way these intuitions are reflected in our framework of M-algebras.

DEFINITION 6.29. Let $\mathcal{A}$ be an M-algebra.. We call $\mathcal{A}$ commutative if all its measurements commute. We call $\mathcal{A}$ classical if all its states are classical. We say that $\mathcal{A}$ is monotonic if $\mathord{\sim}_x$ is monotonic for every $x \in X$.

THEOREM 6.30. *Let $\mathcal{A} = \langle X, M, \rangle$ be a strongly separable Implication M-algebra. Then the following conditions are equivalent:*

- *(i) $\mathcal{A}$ is commutative.*
- *(ii) $\mathcal{A}$ is monotonic.*
- *(iii) $\mathcal{A}$ is classical.*

Proof. We first prove that (i) implies ii). Given any consequence relation $\mathrel{\mid\!\sim}_x$. We need to show that it is monotonic. Assume $\mathrel{\mid\!\sim}_x \beta$. This says $\beta(x) = x$. Let α be any measurement. We then have $\alpha(\beta(x)) = \alpha(x)$. Since α and β commute it follows that $\beta(\alpha(x)) = \alpha(x)$. But this says that $\alpha \mathrel{\mid\!\sim}_x \beta$.

That (ii) implies (iii) can be seen as follows. Given any state x and any measurement α. We need to prove that either $\alpha(x) = x$ or $\alpha(x) = 0$. So assume $\alpha(x) \neq x$. By Proposition 6.19 we have $\mathrel{\mid\!\sim}_x \neg(e_x \rightsquigarrow \alpha)$ and by monotonicity $\alpha \mathrel{\mid\!\sim} \neg(e_x \rightsquigarrow \alpha)$. Hence $\alpha(x) \in FP(\neg(e_x \rightsquigarrow \alpha))$. By lemma 6.17 we have $FP(\neg(e_x \rightsquigarrow \alpha)) = \{x, 0\}$ It follows that $\alpha(x) = 0$.

Clearly, (iii) implies (i).

∎

8 The three faces of truth

Let us, at this point, take the opportunity to intuitively reflect on the three aspects of truth that we naturally encountered in connection with the material presented in this chapter.

There is a familiar aspect of truth that we find in M-algebras. Given a measurement (proposition) α and a state x and assume that $x \in FP(\alpha)$. In this case we say that α *is true* in x. This is *true as being true*. It is in this static manner that truth presents itself to us in traditional logic (and classical physics).

The dynamic view of propositions gives rise to what earlier in this chapter we called *coming true*. A proposition α *comes true* in x if it *is true* in $\alpha(x)$. This is a novelty of quantum logic as we understand it.

Recall the following from chapter 3. Given a physical system $\mathcal{S}$, let H be the Hilbert space associated with $\mathcal{S}$ and let $\mathcal{A}$ be a physical quantity pertaining to $\mathcal{S}$, let μ be an eigenvalue of the Hermitian operator A representing $\mathcal{A}$ and let B be the eigenspace of μ, which is a closed subspace. Given a normed $x \in H$. Then, if we again denote by B the projection on the closed subspace B, the conventional wisdom of quantum mechanics is that the number $|\langle Bx, x \rangle|^2$ represents the probability of getting the value μ

8. THE THREE FACES OF TRUTH

as a result of measuring $\mathcal{A}$ in state x. This is Born's rule. There is, however, a problem with the term probability here. Namely, it has turned out that these 'probabilities' do not permit an interpretation as probabilities in the sense of classical (Kolmogorovian) probability theory. In particular they do not admit a frequency interpretation. This fact has given rise to a whole field of research which has become known as *quantum probability*. Now, we have seen that it could be reasonable to introduce the *coming true* as a logical category which may prove more basic than that of *being true*. It would then be natural to view the 'probabilities' given by Born's rule as the truth values for the coming true. On this view these 'probabilities' have nothing to do with randomness but describe a sort of a tendency of a property — or perhaps more precisely a propensity in the sense of Aristotle — to materialise, i.e. to come true. The connection with (infinite-valued) Lukasiewicz logic is obvious here. One may say that the truth values in Lukasiewicz logic are in fact values for the coming true of a proposition, coming truth values rather than truth values.

The third notion of truth that we naturally encountered in the framework of M-algebras is that of self-referential truth. We will elaborate on this in our study of holistic logics in chapter 8. Recall the following. Given a strongly separable Implication M-algebra $\langle X, M \rangle$ and a state $x \in X$. Then we saw that the state x may be viewed as a logical entity encoding statements about other states and about *itself*. We will see more clearly in chapter 8 that x is a logical entity encoding all true statements *about* itself. The true statements that x encodes are true statements about itself and, in view of the no windows theorems, only about itself. Here truth presents itself as self-referential truth, certainty about oneself so to speak, a sort of self-awareness.

CHAPTER 7

ASPECTS OF QUANTUM REALITY

In this chapter we primarily address those readers who haven't had much contact with quantum mechanics yet. We report on certain typical features of the quantum world which, to these readers, may appear unfamiliar, even strange. Our intention is to convey the impression that quantum mechanics touches on deep issues even beyond the realm of physics.

Quantum mechanics is, in chronological order, the second great revolution of 20th century physics. The first of these two revolutions was of course Einstein's theory of relativity. The theory of relativity forced upon us the revision of long cherished views on space and time. This was a truly profound revision. It seems, however, that quantum mechanics touches on even deeper ground and that it has an even more profound impact on the way we are forced to view the physical world. It seems that one of these issues is that of no less and no more than the nature of (physical) reality itself. This is the topic of a vast body of both seriously scientific and philosophical as well as popular scientific literature.

In fact, it is the issue of reality that constitutes the main reason for including this chapter. This issue is of course of a philosophical, perhaps even of a metaphysical nature, and, since this book is about logic and not about metaphysics, the reader may rightfully ask the question why we want to reflect on a metaphysical problem in a book on modern style logic.

Roughly, the reason is this. We said that we would be searching for logical structures underlying quantum mechanics. Note that the emphasis is on the term 'structure', and, as we said earlier, we are not looking for a new deductive system that could replace classical logic as a tool for reasoning in quantum mechanics.

The logical structures we want to find in quantum mechanics must reflect the way quantum mechanics departs from classical mechanics. We also said hat these structures should make precise, at the logical level, in which sense classical logic is a limiting case of quantum mechanics. The reader, however, can appreciate this only if he has an idea of the phenomena which, at the physical level, are characteristic of quantum mechanics and are typical of the way how quantum mechanics departs from classical mechanics. This is the purpose of this chapter.

1 The wave particle dualism

The first observation to shake our classical view of physical reality was de Broglie's discovery of the wave-particle dualism. Elementary particles have wave nature. An electron for instance can behave like a particle, as we would expect from classical physics, but also as a wave, which is unfamiliar from classical physics. Sometimes the electron behaves particle-like and sometimes it behaves wave-like. De Broglie even discovered a precise mathematical connection between the momentum p of the electron when it 'is' a particle and its wavelength λ when it 'is' a wave. This is the famous de Broglie relation:

$$\lambda = \frac{h}{p}$$

where λ is the wavelength in case of wave-like behaviour and p is the momentum in case of particle behaviour.

Our 'classical world view' suggests the question: Is the electron a particle or a wave? The answer is that this depends on the particular experiment performed in order to observe the electron. There are experiments in which it behaves wave-like and there are experiments in which it behaves particle-like. It thus depends not only on the electron itself what it is but also on the observer. This is what led Bohr to the notion of complementarity. On this view the wave nature and the particle nature are mutually exclusive but *complementary* properties of one and the same physical entity. The reader will probably agree that this phenomenon is hard to reconcile with our traditional way of looking at physical reality.

2 Measurement as an inseparable whole

There is a problem in quantum mechanics with ascribing definite properties to physical systems. Assume we have a quantum system in a certain state and we want to measure a physical quantity or observable E, as is the term in quantum mechanics, say its (total) energy. Then, according to quantum mechanics, this observable may not be sharp. This means that as a result of measurement we may get various values each with a certain probability. Generally, there is a discrete spectrum of values $\lambda_1, \lambda_2, ...$ that E can assume with the corresponding probabilities $p_1, p_2, ...$ such that $\sum_{i=1}^{\infty} p_i = 1$

It is important to note that these probabilities do not describe our ignorance concerning an ensemble from which we 'pick' a property at random. Rather, according to quantum mechanics, the observable E does not have definite values and it is only in the process of measurement that it assumes a certain value with a certain probability.

How can we account for this? Obviously this fact is hardly compatible with the view that there exists an objective physical reality having defi-

2. MEASUREMENT AS AN INSEPARABLE WHOLE

nite preexisting properties. According to the Copenhagen interpretation of quantum mechanics this is an expression of the fact that the quantum world consists of a set of potentialities rather than facts. These potentialities can be actualised in the process of measurement. Heisenberg, one of the chief representatives of the Copenhagen Interpretation, puts it this way: "The atoms or the elementary particles are not as real; they form a world of potentialities or possibilities rather than one of things or facts."

Bohr pointed out that the process of gathering information about the micro-world must, at some point, involve making an experiment in which a laboratory instrument is used. Now, the representatives of the Copenhagen interpretation hold the view that this interaction is an indivisible whole and that it is no longer possible to analyse the parts of the two agents involved, the system to be measured and the measuring instrument in the sense that their 'contribution' to the outcome of the interaction can be described. In Bohr's words, observer and observed system form an "indivisible, unanalysable whole" in the process of observation.

The following is a quotation from Peat's book "Einstein's Moon" [47].[1]

"This holistic view of the nature of the atomic world was the key to the Copenhagen interpretation. It was something totally new in physics, although similar ideas had long been part of Eastern thought and religion. For more than two thousand years, Eastern philosophers had put forward similar views about the unity that lies between the observer and the observed. They had pointed to the illusion of breaking apart a thought from the mind that thinks the thought. Now a similar holism was entering physics.

What then is an electron, a proton or an atom? What properties does a particle have if it only manifests itself in an unanalysable interaction with a piece of apparatus? What does it mean to say that the electron *has* a certain velocity or position if every attempt to measure these properties represent an irreducible act of interference? Indeed it becomes a major problem to speak of the electron as "having" or possessing (definite) properties. And if all the properties of a quantum object become ambiguous, then what sort of reality does it have?

Where then is atomic reality? Heisenberg suggested that the reality now lies in the mathematics. The formalism of matrix mechanics or wave mechanics works perfectly. If you want to know where atomic reality lies, then Heisenberg points to the equations; there is no hope for finding it anywhere else.

Does this mean that here is no reality outside the mathematics? About

---

[1] We are in this section under the influence of Peat's fine book. It is possible that in this section we are — beyond the above quotation — using formulations very similar to those of Peat.

this Bohr is at his most uncompromising. 'There is no quantum world', he says. 'There is only an abstract quantum mechanical description.' "

3 Are there "elements of reality"?

Here we try to give the gist of the EPR-paradox in an informal way. Again we are in this presentation strongly inspired by Peat's book. We give a rough description of the EPR paradox. In 1935, Einstein, Podolsky and Rosen published a now famous paper entitled "Can the quantum mechanical description of physical reality be considered complete?"[15]. In that paper a thought experiment is presented which seemed to shake the newly established edifice of quantum mechanics. This was after the great intellectual struggle between Einstein and Bohr over the foundations of quantum mechanics which was generally believed to have come to an end. But now, after the EPR paper, Einstein had reappeared on the stage with a seemingly devastating argument against quantum mechanics. And, in fact, it took Bohr some time to recover from this blow until he managed to produce a defence against the attack.

We try to present the EPR argument in a nutshell. In this we are aware of the fact that we cannot, in this context, do justice to all its subtleties, let alone Bohr's reply.

Imagine a particle at rest, i.e. with momentum zero, being split into two particles of the same sort, say electrons, moving away in opposite directions with the same velocity. Now, assume particle 1 and particle 2 are far apart from each other and there is no physical interaction between them. We now assume that the position of particle 1 is measured. Once the position of particle 1 is known, the position of particle 2 is known too. So, if we want to know the position of particle 2, it suffices to measure the position of particle 1. Now it is reasonable to assume that performing a measurement on particle 1 does not in any way disturb particle 2. The term used by EPR is "without in any way disturbing". So we can measure the position of particle 2 "without in any way disturbing" it. Put differently, we can predict with certainty the position of particle 2 without in any way disturbing it. Now, EPR say that if a quantity of a system can be predicted with certainty without in any way disturbing it, then this quantity constitutes an "element of reality". So the position of particle 2 constitutes an "element of reality".

The point of the argument is now this. Instead of measuring the position of particle 2 by performing a measurement of the position of particle 1 we may measure the momentum of particle 2 by performing a measurement of momentum on particle 1. So, again we can predict the momentum of particle 2 without in any way disturbing it. Therefore the momentum of particle 2 constitutes an "element of reality". But now note that according to quantum

3. ARE THERE "ELEMENTS OF REALITY"?

mechanics a physical system cannot simultaneously possess both a sharp position and a sharp momentum, which according to EPR are both elements of reality The conclusion EPR now draw is that quantum mechanics does not provide a complete description of reality in that it does not capture all elements of reality. We cannot now go into a detailed discussion of the EPR thought experiment with all its subtleties. But it is obvious from the above that the EPR argument contains at least two tacit assumptions. The first assumption concerns the existence of separate elements of reality, and the second assumption is implicit in the term "without in any way disturbing". The first assumption reflects a picture of a fragmented reality, which consists of separate elements, and the second assumption is what is nowadays called locality. If, therefore, quantum mechanics is a correct description of physical reality, which is widely believed nowadays, then either one or even both of these assumptions must be false. Both assumptions concern the nature of physical reality. So we can say that if quantum mechanics does provide a complete description of physical reality and the EPR argument is thus false, then the EPR argument is false because it rests on a wrong view of physical reality.

But is there any way of proving that this view of reality is wrong? The answer is yes. It is given by *Bell's Theorem*. Bell's theorem provides us with the means of experimentally testing whether physical reality is local or not. This test has been performed in several ingenious experiments by Aspect. And the experimental finding is: *Physical reality is non-local*. Bell's theorem thus permitted us to unveil a feature of physical reality which was undreamed of until it was, on the basis of Bell's theorem, discovered experimentally. This is the reason why Henry Stapp, reputed American physicist and leading expert on the foundations of quantum mechanics, called Bell's Theorem the greatest scientific discovery ever.

What is Bell's theorem? We need not go into the details in order to understand how it opened up the possibility of experimentally testing whether reality is local or not. Bell's theorem is a statement of the form: "If reality is local, then A". The point is that statement A can be *experimentally tested*. Experiment yields: Not A. We conclude: Reality is not local. But there is more to Bell's theorem. Namely, experiment not only yields NOT A, but it yields B. Again, B can be experimentally verified, and B is what quantum mechanics predicts. A triumph of quantum mechanics!

What was Bohr's answer to EPR? Certainly he did not know Bell's theorem yet. But, essentially, Bohr argued that the EPR argument suffered from a fundamental flaw, namely that is was based on a wrong view of (physical) reality. In Bohms words: "He (Bohr) argued that in the quantum domain the procedure by which we analyse classical systems into interacting parts breaks down, for whenever two entities combine to form a single system (even if only for a limited period of time) the process by which they do this is not divisible". The gist of Bohr's reply is that the two particles form an indivisible whole to which our method of fragmenting is not applicable.

4 Bohm on wholeness and his experiment with language

David Bohm's creative life was in its many facets devoted to the puzzle of quantum mechanics. The chain of his thinking contains brilliant ideas all of which became highly influential and form essential parts of the literature on the foundations of quantum mechanics.

His views on physical reality with which he came up after almost half a century of dedicated intellectual work on the problem of understanding quantum mechanics culminated in his book "Wholeness and the Implicate Order" [6]. In that book Bohm consistently argues for a holistic world view in order to account for the puzzles of quantum reality. This holistic world view should replace the fragmenting world view typical of Western thought and of modern Western science such as classical physics. In the first chapter entitled "Fragmentation and Wholeness" he describes the act of observation (measurement) in quantum mechanics as follows: "One can no longer maintain the division between the observer and the observed (which is implicit in the atomistic view that regards each of these as separate aggregates of atoms). Rather, both observer and observed are merging and interpenetrating aspects of one reality, which is indivisible and unanalysable." Another quotation: " ...relativity and quantum theory agree in that they both imply the need to look on the world as an undivided whole, in which all parts of the universe, including the observer and his instruments merge and unite in one totality. In this totality, the atomistic form of insight is a simplification and an abstraction, valid only in some limited context." Bohm then suggests to view reality as an "undivided wholeness in flowing movement". This view put forward by an outstanding representative of modern Western science is undoubtedly reminiscent not only of Eastern thought but also of Heraclitus' philosophy of the world as being in in permanent flux. The metaphor Bohm uses in "Wholeness and the implicate order" is that of a hologram. The word being of Greek origin denotes an instrument 'writing the whole'.

In order to understand the metaphor of the hologram in Bohm's thinking we need to know a bit about the nature of a hologram without, however, having to go into the technical details of actually constructing such an optical instrument. Let us just say this. A hologram is commonly known as a three-dimensional photograph made with the the help of a laser. Using laser light, an interference pattern on a photographic plate is created. The developed film is then illuminated again by laser light. Then a three-dimensional image of the photographed object appears. It is not the impression of three-dimensionality, however, that is most striking about a hologram. Rather it is the following. When we illuminate just a part of the photographic plate, however small, it is, though smaller, still the (whole) image of the *whole object* that appears. So every part of the hologram encodes the whole information possessed by the whole object. Technically, this effect is explained by the wave nature of light and the particular interference effect creating the pattern on the photographic plate.

The main difference between the hologram and the familiar instrument of a *lens* is this. A lens is an optical instrument creating an image of an object in such a way that the parts of the objects correspond to the parts of the image in a one-to-one way. But, due to the wave properties of light, even this is true only approximately. Bohm remarks that thus the case of the lens may be regarded as the limiting case of the hologram. Fragmentation is so to speak the limiting case of wholeness. We may view this intuition as being reflected in the logical framework of chapter 8 as the limiting case theorem.

One of the characteristic feature of the hologram is that it is mirrored (encoded) in all its parts. Our view of reality should, according to Bohm, be holistic in the sense that there is no fragmentation. Rather, the 'parts' are to reflect and encode the 'whole'. This is also strikingly reminiscent of what Leibniz says about the monads, his incorporeal 'atoms of reality'. These monads, Leibniz says, 'mirror' each other. The whole world is thus mirrored in each of its parts.

But for Bohm this is not just a good metaphor. Rather, in the second chapter of "Wholeness and the Implicate Order" entitled "The rheomode — an experiment with language and thought" he takes this picture seriously as the basis for actually constructing new linguistic structures which he thinks are more appropriate to the type of reality suggested by quantum mechanics. What is the rheomode? If our language with is typical subject predicate sentence structure is not the language appropriate for the quantum world, what then does the proper language look like? It would, as Bohm correctly points out, not be practicable to construct a new language having an entirely new structure appropriate for the quantum world. Instead, Bohm proposes a new mode of language similar to that of indicative, imperative, subjunctive etc. He says: ".. will now consider a mode in which movement is to be taken as primary in our thinking and in which this motion will be incorporated into the language structure by allowing the verb rather than the noun to play a primary role... For the sake of convenience we will give this mode a name, i.e. the rheomode (rheo is from a Greek verb, meaning to flow). At least in the first instance the rheomode will be an experiment with language, concerned mainly with trying to find out whether it is possible to create a new structure that is not so prone towards fragmentation as is the present one."

Here Bohm proposes to consider verbs such as 'to levate' or 'relevate', which are not part of language so far or any longer, because they may have dropped out of language. Why does he choose the notion of relevance as the starting point for his inquiry into the rheomode? When using the term 'relevant' we focus on a whole bunch of things. First, clearly, we focus on something we consider relevant, say a sentence uttered in a discussion. But we also focus on the process of thought or perception as a result of which this sentence appears relevant. Moreover, its relevance depends on the context, which can change and does change. So, in making statements about relevance we refer to an integrated whole involving language, perception, extra-linguistic contexts, and this whole is in flux. The boundaries between relevance and irrelevance are not sharp as the structure of language with the dominating nouns 'relevance' and 'irrelevance' would suggest. This, Bohm says, is a case for the rheomode. He proposes to introduce the verb 'to relevate' into the language which, he says, should mean " to lift a certain context into attention again, for a particular context as indicated by thought and language". The term 're-levant' then denotes the state of being 're-levated'. Bohm says: "So when relevance or irrelevance is communicated, one has to understand that this is not a hard and fast division between opposing categories but rather, an expression of an ever-changing perception, in which it is possible, for the moment, to see a fit or non-fit between the content lifted into attention and the context to which it refers."

5 Experimenting with logic?

The reader may ask the question why we dwelled so long on Bohm's experiment with language. What reason is there for describing it in this book? The reason is that we could have taken Bohm's book as an inspiration for performing another experiment in order to deal with the issue of reality in quantum mechanics. This experiment would not have been with language, as is Bohm's, but with logic, which is the topic of this book. Bohm's idea was that the language we use in everyday life and also in classical physics is not appropriate for quantum mechanics because the fragmenting view of reality underlying classical physics needs revision in quantum mechanics and that — for quantum mechanics — we need a language reflecting the holistic features of quantum reality. In sketching a new mode of language, the rheomode, Bohm tried to show how, in principle, such a language could

5. EXPERIMENTING WITH LOGIC?

look like.

Now, clearly, there is the analogous consideration with logic in place of language. Our fragmenting world view underlying classical mechanics is not only reflected in language but also in logic. It would, therefore, be equally desirable to construct a logic that mirrors the holistic features of quantum reality. It seems, however, that this is more easily said than done. How could this work?

We could take the following consideration as a starting point. One of the fundamental differences between classical and quantum mechanics undoubtedly concerns the role of observation. In classical physics observation or, say, measurement is a sort of 'looking' at a certain objective reality outside the observer. It is taken for granted that there exists a clear cut separation between the observer and the reality observed. This dualistic picture is hard to defend in quantum mechanics. In Bohr's and also in Bohm's words, the process of measurement is "unanalysable". In the process of measurement in quantum mechanics the observer and the observed system form an inseparable whole. The dualistic view of the observer on the one hand and the observed reality on the other, of the subject on the one hand and the object on the other, seems to be untenable in quantum mechanics.

In modern logic we have, since Tarski, a similar dualism which is not even confined to classical logic. It concerns the separation between the two components of a modern logical system. Such a system, generally, displays the separation between syntactic representation on the hand and semantic representation on the other. In Girard's words, this opposition between *semantics* (the world) and *syntax* (its representation), which underlies logical realism, is highly problematic. It seems that it is one of the objectives of Girard's 'ludics' (see [25]) as well as of what has become known as game-theoretic semantics in recent years to overcome this dualism in modern logic.

In a modern-style logical system the issue of reality enters the stage via semantics. It is in the choice of the semantic structures for a logic that we make a certain commitment to our view of the structure of reality. One of the functions of the semantic structures is to define the notion of truth of a formula in these structures. Take for instance predicate logic. Given an (atomic) formula, i.e. a syntactic entity, of the form $P(c)$, where P is a unary predicate symbol and c is an individual constant symbol. The intended meaning of $P(c)$ is of course 'c has property P'. In the semantics of predicate logic the above formula is treated as follows. Given a language of predicate logic. Such a language has certain relation symbols. The relations of arity 0 are called individual constant symbols. The semantic structures for such a language are so called relational structures. A relational structure is a set X together with a family of relations on X. For each relation symbol R we have a relation of the corresponding arity in the relational structure serving as the denotation of R. In the above example the denotation of the symbol P is a subset of X say again denoted by P and the denotation of the individual constant c is an element of X again denoted by c. The truth conditions for the formula $P(c)$ is that $P(c)$ is true iff $c \in P$.

If we now claim that the logic semantically presented in the style described is not just a mathematical construction but constitutes a tool for actually reasoning about the world, then we have made a commitment to our view of reality. Namely, we have committed ourselves to the view that the relational structures we chose as our semantic structures properly represent or reflect reality, at least the type of reality we reason about in predicate logic. We commit ourselves to the view that there exist individuals, that these individuals have properties represented as predicates, that there exist relations between individuals and so on. This amounts to the formal reconstruction of the fragmenting world view which needs revision in the quantum domain.

In view of the notorious problem of reality in quantum mechanics and in view of our ignorance of the nature of that reality we may propose an experiment with the following characteristic:

Construct logical systems with a minimum of commitment to the structure of reality.

What does this mean? First we have to be a bit more precise regarding the term reality and reformulate the task as constructing logical systems with a minimum of commitment to the structure of reality external to the logic. Since we can hardly deny the logic to be real, the following is an immediate consequence.

The only semantic structures relevant to the logics to be constructed are the logics themselves.

If we think of a logical system as a system of reasoning about some domain of reality, then for a system satisfying the above principle, this domain of reality is nothing but the logical system itself. Thus the only (domain of) reality such a logic can 'reason' about is itself. Such systems — yet to be constructed — reason about themselves only, about their own 'internal working' so to speak. This would indeed be a radical way of getting rid of the problem of reality in logic and also of the dualism between syntactic and semantic representation.

The above principle has several implications.

The first implication concerns the notion of truth. Generally, in the spirit of the correspondence theory of truth, the notion of truth of a formula in some logical language is defined as 'truth in a model', i.e. relative to the

5. EXPERIMENTING WITH LOGIC?

semantic structures for the logic. In view of the above principle our notion of truth will be a notion of *self-referential truth*.

We may further ask the question what formulas can be true or false in view of the above principle. Obviously these are the formulas that 'talk' about the logic, formulas making a statement about the logic, i.e. meta-statements. This poses another problem if we are to make sense of the above principle. Namely, if meta-statements are the only formulas that can be true or false, then the object language must be capable of expressing meta-statements. Moreover, we must require it to be rich in expressive power with respect to meta-statements, i.e. it must be able to express *all* statements that can be a made in the language which may reasonably be called *the* metalanguage of the logic. We thus expect the object language to contain the metalanguage of the logic.

And, if we require the logic, as usual, to be sound and complete, this means that we require it to be capable of proving all true meta-statements and only those. And thus, if we require soundness, we require *self-referential soundness* and if we require completeness, we require *self-referential completeness*.

What do we expect to be the logic of the meta-statements? Since the meta-statement 'talk' about some reality, namely the logic itself, and we, in this, assume the correspondence theory of truth, we expect them to obey classical logic. This means that the logic of the metalanguage must be classical logic.

We have not actually performed the hypothetical experiment described. Rather, in the next chapter, we describe what could have been its outcome. Namely, in the next chapter we introduce structures which we call *holistic logics* and which satisfy the conditions described.

CHAPTER 8

HOLISTIC LOGICS

In this chapter we introduce, as in chapter 5, certain structures which are abstractions from Hilbert space. In chapter 5 the focus was on propositions. The structures we introduced, namely M-algebras, were designed to capture the way how propositions act on states. States were, there, primitive notions. In chapter 6 we went one step further. We focused on the nature of the states themselves. In all this, however, we did not leave the framework of M-algebras. In this chapter we also abstract from this. The central concept we introduce in this chapter is that of a *holistic logic*. Our motivation for choosing this term is given by our intuitive reflections on the concept of a state in a dynamic logical framework, see chapter 5. We saw that states in such a framework must encode other states. This intuition is well reflected in the logical structures we call holistic logics. Loosely speaking, in such systems "everything is encoded in (almost) everything".

There is a certain overlap between this chapter and the theory of strongly separable Implication M-algebras of chapter 8 which, in a unified framework, could be avoided. This deliberate redundancy permits the reader to study chapters 6 and 8 independently.

Throughout this chapter we assume the language of propositional logic as described in chapter 1.

1 Consequence revision systems

1.1 Formal motivation: the Lindenbaum algebra viewed as an operator algebra

In order to motivate the concepts we are going to introduce we start with an observation from classical logic. We denote the language by Fml. Recall that $\vdash$ denotes the consequence relation of classical logic. Given a formula α, we may form a new consequence relation $\vdash_\alpha$ as follows: $\beta \vdash_\alpha \gamma$ iff $\alpha \wedge \beta \vdash \gamma$. We get a class of consequence relations $\mathcal{C} = \{\vdash_\alpha | \ \alpha \in Fml\}$. By the deduction theorem of classical logic we have $\beta \vdash_\alpha \gamma$ iff $\vdash_\alpha \beta \to \gamma$ for all $\vdash_\alpha \in \mathcal{C}$. We say that $\to$, i.e. material implication, is an internalising connective for $\mathcal{C}$. Again, given $\alpha \in Fml$ and $\vdash_\beta \in \mathcal{C}$, we may form the consequence relation $\vdash_{\alpha \wedge \beta}$. Thus every $\alpha \in Fml$ induces an operator $\overline{\alpha} : \mathcal{C} \to \mathcal{C}$. We have $\overline{\alpha} = \overline{\beta}$ iff α and β are classically equivalent. It is readily verified that the class of operators is partially ordered by: $\overline{\alpha} \leq \overline{\beta}$ iff $\alpha \vdash \beta$. Moreover, it is routine to verify that this structure forms a Boolean algebra isomorphic to the Lindenbaum algebra of classical logic. Observe that $\overline{\beta\alpha} = \overline{\alpha}$ iff $\alpha \vdash \beta$. This is our motivating example of what we will call a *consequence revision system*. Its main ingredients are a *class of consequence relations* $\mathcal{C}$, a function $F : Fml \times \mathcal{C} \to \mathcal{C}$ and a connective which is an internalising connective for all consequence relations of $\mathcal{C}$. In this case this is material implication. We have $\vdash_\alpha \beta \to \gamma$ iff $\beta \vdash_\alpha \gamma$ for any α. The structure of interest is the triple $\mathcal{L} = \langle \mathcal{C}, \mathcal{F}, \to \rangle$.

There is a straightforward generalisation of the above consideration. We could have started with any consistent set of formulas Σ and the consequence relation $\vdash_\Sigma$ defined by: $\alpha \vdash_\Sigma \beta$ iff $\Sigma \cup \{\alpha\} \vdash \beta$ and would by the same procedure as above have arrived at the structure $\mathcal{L}_\Sigma = \langle \mathcal{C}_\Sigma, \mathcal{F}_\Sigma, \to \rangle$. Note that, by the deduction theorem of classical logic, material implication is still the internalising connective in this more general case.

1.2 The concept of a consequence revision system

We will, in this chapter, be concerned with classes of consequence relations and must therefore consider conditions these consequence relations are supposed to satisfy. These conditions go beyond those stated in chapter 1. But one should note that all these conditions hold in Hilbert space.

We denote the universal (inconsistent) universal consequence relation by 0. We assume that for the consequence relations we consider this is equivalent to the existence of a formula α such that $\mathrel{|\!\sim} \alpha$ and $\mathrel{|\!\sim} \neg\alpha$. Any class of consequence relations considered is assumed to contain 0. That means we assume for any $\mathrel{|\!\sim} \neq 0$ that for no $\alpha \in Fml$ we have $\mathrel{|\!\sim} \alpha$ and $\mathrel{|\!\sim} \neg\alpha$.

Given a class $\mathcal{C}$ of consequence relations. Then we write $\alpha \mathrel{|\!\sim}_\mathcal{C} \beta$ iff $\alpha \mathrel{|\!\sim} \beta$ for every $\mathrel{|\!\sim} \in \mathcal{C}$. We say $\alpha \equiv_\mathcal{C} \beta$ if $\alpha \mathrel{|\!\sim}_\mathcal{C} \beta$ and $\beta \mathrel{|\!\sim}_\mathcal{C} \alpha$.

Minimal Conditions 1

Let us for the sake of convenience again mention here the minimal conditions of chapter 1 which are generally imposed on consequence relations.

Reflexivity
$$\alpha \mathrel{|\!\sim} \alpha$$

Cut
$$\frac{\alpha \wedge \beta \mathrel{|\!\sim} \gamma,\ \alpha \mathrel{|\!\sim} \beta}{\alpha \mathrel{|\!\sim} \gamma}$$

Restricted Monotonicity
$$\frac{\alpha \mathrel{|\!\sim} \beta,\ \alpha \mathrel{|\!\sim} \gamma}{\alpha \wedge \beta \mathrel{|\!\sim} \gamma}$$

1. CONSEQUENCE REVISION SYSTEMS

For a given consequence relation $\mid\sim$ define

$$\alpha \equiv \beta \text{ iff } \alpha \mid\sim \beta \text{ and } \beta \mid\sim \alpha$$

Minimal Conditions 2

Moreover, we impose the following conditions on the special consequence relations studied in this book.

$$\alpha \equiv \neg\neg\alpha$$

$$\top \equiv \alpha \vee \neg\alpha$$

$$\bot \equiv \alpha \wedge \neg\alpha$$

$$\alpha \wedge \beta \mid\sim \alpha$$

$$\mid\sim \alpha \text{ and } \mid\sim \beta \text{ implies } \mid\sim \alpha \wedge \beta$$

$$\alpha \wedge \beta \mid\sim \beta$$

$$\alpha \mid\sim \alpha \vee \beta$$

$$\beta \mid\sim \alpha \vee \beta$$

$$\mid\sim \alpha \vee \neg\alpha$$

$$\alpha \mid\sim \top$$

$$\bot \mid\sim \alpha$$

$$\neg(\alpha \wedge \beta) \equiv \neg\alpha \vee \neg\beta$$

$$\neg(\alpha \vee \beta) \equiv \neg\alpha \wedge \neg\beta$$

Note that we have from 2.3 of chapter 1 $\mid\sim \alpha \wedge \beta$ if $\mid\sim \alpha$ and $\mid\sim \beta$.

The conditions we imposed so far are 'local' in the sense that they are imposed separately on every single consequence relation belonging to the class considered. We, moreover, impose the following conditions which have a global character in the sense that they are related to the class $\mathcal{C}$ as a whole.

$$\frac{\alpha \mid\sim_C \gamma, \beta \mid\sim_C \gamma}{\alpha \vee \beta \mid\sim_C \gamma}$$

$$\frac{\alpha \mid\sim_C \beta}{\neg\beta \mid\sim_C \neg\alpha}$$

The reason for imposing these conditions is that we want to have the algebraic structures arising from these logical structures to have certain desirable properties that are actually fulfilled in the case of the concrete structures arising in connection with quantum mechanics. So, in contrast to two chapters 5 and 6 we want the relevant algebraic structures to be lattices. This brings us closer to Hilbert space, where the algebraic structures relevant from the logical point of view are in fact lattices.

Let us n w define the key concept of a consequence revision system.

DEFINITION 8.1. Let $\mathcal{C}$ be a class of consequence relations over Fml satisfying the conditions described. Let F be a function

$$F : Fml \times \mathcal{C} \to \mathcal{C}.$$

We say that F is an action on $\mathcal{C}$ iff for every $\mathrel{\mid\!\sim} \in \mathcal{C}$ and $\alpha, \beta \in Fml$ the following conditions are satisfied.

(i) $F(\top, \mathrel{\mid\!\sim}) = \mathrel{\mid\!\sim}$

(ii) $F(\alpha, \mathrel{\mid\!\sim}) = 0$ iff $\mathrel{\mid\!\sim} \neg \alpha$

(iii) $F(\beta, F(\alpha, \mathrel{\mid\!\sim})) = F(\alpha, \mathrel{\mid\!\sim})$ iff $\alpha \mathrel{\mid\!\sim} \beta$

If F is an action on $\mathcal{C}$, we call the pair $\langle \mathcal{C}, F \rangle$ a *consequence revision system* (CRS).

Note that by $\mathrel{\mid\!\sim} \alpha$ we mean $\top \mathrel{\mid\!\sim} \alpha$. For a given class $\mathcal{C}$ of consequence relations call the formulas α and β $\mathcal{C}$-equivalent, in symbols $\alpha \equiv_\mathcal{C} \beta$, if for every $\mathrel{\mid\!\sim} \in \mathcal{C}$ we have $\alpha \mathrel{\mid\!\sim} \beta$ and $\beta \mathrel{\mid\!\sim} \alpha$.

Remark: We are aware of the fact that the way we use the term revision in the above definition does not fully capture the way it is used in traditional revision theory (see for instance [1]). If at all, the action of formulas on consequence relations as defined above represents a simple type of revision. Condition (*ii*) above says that given a consequence relation $\mathrel{\mid\!\sim}$ and a formula α which is inconsistent with $\mathrel{\mid\!\sim}$ then the result of 'revising' $\mathrel{\mid\!\sim}$ by α is the inconsistent consequence relation. The corresponding case in traditional revision theory is that of a theory T and a formula α inconsistent with T. The result of revising T by α usually denoted by $T * \alpha$ is then, according to traditional revision theory, not necessarily the inconsistent theory. Since, however, in our most important examples, namely those arising from Hilbert spaces, we are concerned with a process which, in the intuitive sense, deserves to be called revision, we freely use the term revision. Every $\alpha \in Fml$ induces a (revision) operator on $\mathcal{C}$

$$\overline{\alpha} : \mathcal{C} \to \mathcal{C}$$

1. CONSEQUENCE REVISION SYSTEMS

via
$$\overline{\alpha}\,\mid\!\sim\; =:\; F(\alpha,\mid\!\sim)$$

For $\overline{\alpha}\,\mid\!\sim$ we will also write $\mid\!\sim_\alpha$.

Denote the class of these operators by $\overline{Fml}$. We have $\overline{\alpha} = \overline{\beta}$ iff $\alpha \equiv_\mathcal{C} \beta$.

LEMMA 8.2. *For any $\alpha \in Fml$ we have $\overline{\alpha} \circ \alpha = \overline{\alpha}$.*

Proof. By *Reflexivity* we have $\alpha \mid\!\sim \alpha$ for every $\mid\!\sim\, \in \mathcal{C}$. Then the claim follows by condition (iii) of the definition of an action. ∎

LEMMA 8.3. *Let $\langle \mathcal{C}, F \rangle$ be a CRS. Then for any $\mid\!\sim\, \in \mathcal{C}$ the following conditions are equivalent*

- (i) $\mid\!\sim \alpha$
- (ii) $\mid\!\sim_\alpha \,=\, \mid\!\sim$
- (iii) *There exists a $\mid\!\sim_1 \in \mathcal{C}$ such that $\mid\!\sim_{1,\alpha}\,=\,\mid\!\sim$*

Proof. For the equivalence of (i) and (ii) observe first that $\mid\!\sim_{\top,\alpha}\,=\,\mid\!\sim_\alpha$. By condition (iii) of the definition of an action we have that $\top \mid\!\sim \alpha$ iff $\mid\!\sim_\alpha\,=\,\mid\!\sim_{\top,\alpha}\,=\,\mid\!\sim_\top\,=\,\mid\!\sim$. Clearly, (ii) implies (iii). In order to show that (iii) implies (ii) suppose $\mid\!\sim_{1,\alpha}\,=\,\mid\!\sim$. Note that by *Reflexivity* we have $\alpha \mid\!\sim_1 \alpha$. Then it follows by condition (iii) of the definition of an action that $\mid\!\sim_\alpha\,=\,\mid\!\sim$. ∎

LEMMA 8.4. *$\alpha \mid\!\sim \beta$ if $\mid\!\sim_\alpha \beta$.*

Proof. Suppose $\alpha \mid\!\sim \beta$. By condition (iii) of the definition of an action this is equivalent to $\mid\!\sim_{\alpha,\beta}\,=\,\mid\!\sim_\alpha$. By (i) of the above lemma this means that $\mid\!\sim_\alpha \beta$. ∎

It follows by the above two lemmas that $\mid\!\sim_\alpha\,=\,\mid\!\sim_\beta$ implies $\alpha \equiv \beta$, i.e. $\alpha \mid\!\sim \beta$ and $\beta \mid\!\sim \alpha$. We see that $\overline{\alpha} = \overline{\beta}$ iff $\alpha \equiv_\mathcal{C} \beta$.

DEFINITION 8.5. *Let $\langle \mathcal{C}, F \rangle$ be a CRS. Then define the proposition $[\alpha]$ by*

$$[\alpha] =:\; \{\mid\!\sim \;\mid\; \mid\!\sim \alpha\}$$

We denote the class of propositions of $\langle \mathcal{C}, F \rangle$ by Prop.

It is routine to verify the statements made in the following lemma.

LEMMA 8.6. *Let $\langle \mathcal{C}, F \rangle$ be a CRS. Then*

$$\overline{\alpha} \leq \overline{\beta} \text{ iff } [\alpha] \subset [\beta]$$

$$\overline{\alpha} = \overline{\beta} \text{ iff } [\alpha] = [\beta]$$

$$\alpha \mathrel{\vdash}_\mathcal{C} \beta \text{ iff } [\alpha] \subset [\beta]$$

$$\alpha \equiv_\mathcal{C} \beta \text{ iff } [\alpha] = [\beta]$$

$$\overline{\alpha} \leq \overline{\beta} \text{ iff } \overline{\neg \beta} \leq \overline{\neg \alpha}$$

$$[\alpha] \subset [\beta] \text{ iff } [\neg \beta] \subset [\neg \alpha]$$

The conditions we imposed on the consequence relations guarantee that the following holds.

PROPOSITION 8.7. *For any CRS both $\langle \overline{Fml}, \leq \rangle$ and $\langle Prop, \subset \rangle$ are lattices. For $\overline{\alpha}, \overline{\beta} \in \overline{Fml}$ and $[\alpha], [\beta] \in Prop$ the greatest lower bounds are $\overline{\alpha \wedge \beta}$ and $[\alpha \wedge \beta]$ respectively. The lowest upper bounds are given by $\overline{\alpha \vee \beta}$ and $[\alpha \vee \beta]$ respectively. The unit and the zero element are given by $[\top]$ and $[\bot]$ respectively.*

Given a CRS $\langle \mathcal{C}, F \rangle$. Then define unary operations $^* : \overline{Fml} \to \overline{Fml}$ and $^* : Prop \to Prop$ as follows.

$$\overline{\alpha}^* =: \overline{\neg \alpha}$$

and

$$[\alpha]^* =: [\neg \alpha]$$

Note that in view of lemma 8.6 these operations are well defined. Moreover, we define a mapping $\psi : \overline{Fml} \to Prop$ by

$$\psi(\overline{\alpha}) = [\alpha]$$

again, by Lemma 8.6 this mapping is well defined. It is routine to verify the following proposition which bears an analogy to the well known fact that in Hilbert space the lattice of projections and the lattices of closed subspaces are isomorphic (orthomodular) lattices.

PROPOSITION 8.8. *Let $\langle \mathcal{C}, F \rangle$ be a CRS. Then*

- *$\langle Fml, \leq, ^* \rangle$ and $\langle Prop, \subset, ^* \rangle$ are ortholattices.*

- *ψ is an isomorphism between ortholattices.*

1. CONSEQUENCE REVISION SYSTEMS

We now define the concept of an *internalising connective*, which is essentially already familiar from Chapter 6. From the logical point of view the concept of an internalising connective is the link between the object level and the meta level. Whenever we use the term connective we mean a connective definable by the usual propositional connectives in the following sense. We say that $\alpha \leadsto \beta$ is definable if there exists a formula of propositional logic $\varphi(p,q)$ with exactly two propositional variables such that the formula $\alpha \leadsto \beta$ is the result of uniformly substituting α in φ for p and β for q. We say that φ defines $\leadsto$. Given two connectives $\leadsto_1$ and $\leadsto_2$ defined by φ_1 and φ_2 respectively. Then we say that $\leadsto_1$ and $\leadsto_2$ are classically equivalent if φ_1 and φ_2 are classically equivalent. Consider for instance the Sasaki hook $\leadsto_s$ which is defined by $\varphi(p,q) =: \neg p \vee (p \wedge q)$. This says that $\alpha \leadsto_s \beta$ is just short for $\neg \alpha \vee (\alpha \wedge \beta)$. The Sasaki hook is classically equivalent to material implication.

DEFINITION 8.9. Let $\vdash$ be a consequence relation and $\leadsto$ a connective such that $\alpha \vdash \beta$ iff $\vdash \alpha \leadsto \beta$. Then we say that $\leadsto$ is an internalising connective for $\vdash$. Given a CRS $\langle \mathcal{C}, F \rangle$. Then we say that $\leadsto$ is an internalising connective for $\langle \mathcal{C}, F \rangle$ iff $\leadsto$ is an internalising connective for all $\vdash \in \mathcal{C}$.

PROPOSITION 8.10. *Let $\langle \mathcal{C}, F \rangle$ be a CRS and let $\leadsto$ be an internalising connective for $\langle \mathcal{C}, F \rangle$. Then the following holds.*

- *(i) $\alpha \vdash (\beta \leadsto \gamma)$ iff $\beta \vdash_\alpha \gamma$*
- *(ii) $\{\vdash \mid \alpha \vdash \beta\}$ is a proposition, namely $[\alpha \leadsto \beta]$*

Proof. By lemma 8.4 we have $\alpha \vdash (\beta \leadsto \gamma)$ iff $\vdash_\alpha (\beta \leadsto \gamma)$. Since $\leadsto$ is internalising, this is equivalent to $\beta \vdash_\alpha \gamma$. This proves (i).
(ii) follows from the fact that $\leadsto$ is internalising. ∎

Note that in view of the above we can in case we have an internalising connective $\leadsto$ describe the process of revision simply as follows. Revise the consequence relation $\vdash$ by α so as to get $\vdash_\alpha$. Then γ can be proved from β in $\vdash_\alpha$ iff $\beta \leadsto \gamma$ can be proved from α in $\vdash$.

Given a class of consequence relations $\mathcal{C}$ and two connectives $\leadsto_1$ and $\leadsto_2$. We then say that $\leadsto_1$ and $\leadsto_2$ are $\mathcal{C}$-equivalent iff for all formulas $\alpha, \beta \in Fml$ we have $\alpha \leadsto_1 \beta \equiv_\mathcal{C} \alpha \leadsto_2 \beta$.

LEMMA 8.11. *Let $\langle \mathcal{C}, F \rangle$ be a CRS. Then any two internalising connectives for $\langle \mathcal{C}, F \rangle$ are $\mathcal{C}$-equivalent.*

Proof. Let $\leadsto_1$ and $\leadsto_2$ be two internalising connectives for $\langle \mathcal{C}, F \rangle$. By symmetry it suffices to prove that $\alpha \leadsto_1 \beta \vdash_\mathcal{C} \alpha \leadsto_2 \beta$. So let $\vdash$ be any element of $\mathcal{C}$ such that $\mathrel{\vdash} \alpha \leadsto_1 \beta$. Since $\leadsto_1$ is internalising, we have $\alpha \mathrel{\vdash} \beta$ and, since $\leadsto_2$ is internalising, $\mathrel{\vdash} \alpha \leadsto_2 \beta$. ∎

The above lemma says that the action 'determines' the internalising connective modulo $\mathcal{C}$-equivalence. The next lemma states a sort of converse for this, namely that the internalising connective 'determines' the action.

LEMMA 8.12. *Let $\langle \mathcal{C}, F_1 \rangle$ and $\langle \mathcal{C}, F_2 \rangle$ be CRS and let $\leadsto$ be a connective which is internalising for both. Then we have $F_1 = F_2$.*

The concept of an action of formulas on a class of consequence relations can serve as a vehicle for studying the interplay between properties of the operator algebra on the one hand and properties of the class of consequence relations on the other. Generally, properties of the former type are algebraic in nature, whereas properties of the latter type are logical in nature. The orthocomplemented lattice of operators and thus the lattice of propositions may have the algebraic property of being orthomodular and we may ask the question what is the 'logical' counterpart of that algebraic property. This is a situation familiar from various branches of mathematics.

The concept of an orthomodular lattice is a dominant concept in virtually all approaches to quantum logic. It is so to speak the quantum logical counterpart of the concept of a Boolean algebra in classical logic. This fact is highlighted by the following theorem.

THEOREM 8.13. *Let $\langle \mathcal{C}, F \rangle$ be a CRS such that for any $\mathrel{\vdash} \in \mathcal{C}$, $\mathrel{\vdash} \alpha \leadsto_s \beta$ implies $\alpha \mathrel{\vdash} \beta$ and let $\leadsto$ be an internalising connective for $\langle \mathcal{C}, F \rangle$. Then $\langle \overline{Fml}, \leq, ^* \rangle$ and thus $\langle Prop, \subset, ^* \rangle$ are orthomodular lattices and $\leadsto$ is $\mathcal{C}$-equivalent to $\leadsto_s$.*
If $\leadsto_s$ is $\mathcal{C}$-equivalent to $\to$, i.e. material implication, then the above lattices are Boolean algebras.

Proof. In view of proposition 8.8 it suffices to prove orthomodularity. We first show that for any $\mathrel{\vdash} \in \mathcal{C}$

$$(1) \quad \alpha \wedge (\alpha \leadsto \beta) \mathrel{\vdash} \beta$$

By Lemma 8.4 it suffices to show that $\mathrel{\vdash}_{\alpha \wedge (\alpha \leadsto \beta)} \beta$. By the same lemma we get $\mathrel{\vdash}_{\alpha \wedge (\alpha \leadsto \beta)} \alpha \wedge (\alpha \leadsto \beta)$ and thus $\mathrel{\vdash}_{\alpha \wedge (\alpha \leadsto \beta)} \alpha$ and $\mathrel{\vdash}_{\alpha \wedge (\alpha \leadsto \beta)} \alpha \leadsto \beta$. Moreover, since $\leadsto$ is internalising, we have $\mathrel{\vdash}_{\alpha \wedge (\alpha \leadsto \beta), \alpha} \beta$. But $\mathrel{\vdash}_{\alpha \wedge (\alpha \leadsto \beta), \alpha} = \mathrel{\vdash}_{\alpha \wedge (\alpha \leadsto \beta)}$, since $\mathrel{\vdash}_{\alpha \wedge (\alpha \leadsto \beta)} \alpha$. Now (1) is proved.
It follows that

$$(2) \quad \overline{\alpha} \wedge \overline{\alpha \leadsto \beta} \leq \overline{\beta}$$

We now prove that the operator $\overline{\alpha \leadsto \beta}$ has the following property.

1. CONSEQUENCE REVISION SYSTEMS

(3) $\overline{\alpha} \wedge \overline{\beta} \leq \overline{\gamma}$ implies $\overline{\alpha \leadsto_s \beta} \leq \overline{\alpha \leadsto \gamma}$.

For this we have to use that every $\mathord{\vdash} \in \mathcal{C}$ satisfies *Cut*. Assume $\overline{\alpha} \wedge \overline{\beta} \leq \overline{\gamma}$ and let $\mathord{\vdash} \in \mathcal{C}$ be such that $\mathord{\vdash} \alpha \leadsto \beta$. We then have $\alpha \wedge \beta \mathrel{\vdash} \gamma$ and, since $\leadsto$ is internalising, $\alpha \mathrel{\vdash} \beta$. Then we get, using *Cut*, that $\alpha \mathrel{\vdash} \gamma$ and again, since $\leadsto$ is internalising, $\mathord{\vdash} (\alpha \leadsto \gamma)$. Thus $\overline{\alpha \leadsto \beta} \leq \overline{\alpha \leadsto \gamma}$.
Now, by the hypothesis, $\mathord{\vdash} (\alpha \leadsto_s \beta)$ implies $\alpha \mathrel{\vdash} \beta$ and thus, since $\leadsto$ is internalising, $\mathord{\vdash} (\alpha \leadsto \beta)$. This means $\overline{\alpha \leadsto_s \beta} \leq \overline{\alpha \leadsto \beta}$. By transitivity we have $\overline{\alpha \leadsto_s \beta} \leq \overline{\alpha \leadsto \gamma}$.
We have proved that, if $\overline{\alpha} \wedge \overline{\beta} \leq \overline{\gamma}$, then $\mathord{\vdash} \alpha \leadsto_s \beta$ implies $\mathord{\vdash} \alpha \leadsto \gamma$ for any $\mathord{\vdash} \in \mathcal{C}$, which means $\overline{\alpha \leadsto_s \beta} \leq \overline{\alpha \leadsto \gamma}$. We now get by (2), (3) and Mittelstaedt's theorem 2.23 that $\langle \overline{Fml}, \leq, ^* \rangle$ and thus $\langle Prop, \subset, ^* \rangle$ are orthomodular and, moreover, $\overline{\alpha \leadsto \beta} = \overline{\alpha}^* \vee (\overline{\alpha} \wedge \overline{\beta})$. From this it follows that $\leadsto$ and $\leadsto_s$ are $\mathcal{C}$-equivalent.
That the lattices under consideration are Boolean if $\leadsto_s$ is $\mathcal{C}$-equivalent to material implication, again, follows by Mittelstaedt's theorem. ∎

Remark: Note that in the above proof two 'logical' properties of the class $\mathcal{C}$ play a crucial role in establishing the fact that the lattices $\langle \overline{Fml}, \leq, ^* \rangle$ and $\langle Prop, \subset \ ^* \rangle$ have the algebraic property of being orthomodular. The first 'logical' property is that an internalising connective having a certain property exists for $\langle \mathcal{C}, F \rangle$. This property of an action can, as we will see, be viewed as a generalisation of the property that the deduction theorem holds. The second crucial property is that all consequence relations of $\mathcal{C}$ satisfy *Cut*.
For the purposes of this paper we introduce the following notion of a *logic*.

DEFINITION 8.14. Let $\langle \mathcal{C}, F \rangle$ be a *CRS* and $\leadsto$ an internalising connective for $\langle \mathcal{C}, F \rangle$. Then call the triple $\mathcal{L} = \langle \mathcal{C}, F, \leadsto \rangle$ a logic.

We may thus interpret the above theorem as essentially saying that for a *CRS* to become a logic (with $\leadsto_s$ as its internalising connective), it is necessary that the lattice of operators $\langle \overline{Fml}, \leq, ^* \rangle$ and thus the lattice of propositions $\langle Prop, \subset, ^* \rangle$ have the algebraic property of being orthomodular. Given a consequence relation $\mathord{\vdash}$, then define $C(\mathord{\vdash}) =: \{\alpha \mid \mathord{\vdash} \alpha\}$. We have the

PROPOSITION 8.15. *Let* $\mathcal{L} = \langle \mathcal{C}, F, \leadsto \rangle$ *be a logic. Given* $\mathord{\vdash}_1, \mathord{\vdash}_2 \in \mathcal{C}$. *Then* $C(\mathord{\vdash}_1) = C(\mathord{\vdash}_2)$ *iff* $\mathord{\vdash}_1 = \mathord{\vdash}_2$.

Proof. Suppose $C(\mathord{\vdash}_1) = C(\mathord{\vdash}_2)$ and let $\alpha \mathrel{\vdash}_1 \beta$. It follows, since $\leadsto$ is internalising that $\mathord{\vdash}_1 (\alpha \leadsto \beta)$ and thus by the hypothesis $\mathord{\vdash}_2 (\alpha \leadsto \beta)$. Again, since $\leadsto$ is internalising, we get $\alpha \mathrel{\vdash}_2 \beta$, thus $\mathord{\vdash}_1 \subset \mathord{\vdash}_2$. By symmetry we also get the other inclusion. ∎

1.3 Classical logic revisited

Let us now return to our motivating example from classical logic and look at it from the point of view of the framework developed in the last subsection. Let $\vdash$ denote classical consequence and let $\Sigma \subset Fml$ be any consistent set of formulas. Define the class $\mathcal{C}_{\Sigma,\alpha}$ of consequence relations as follows. For a given formula α, define $\vdash_{\Sigma,\alpha}$ by:

$$\beta \vdash_{\Sigma,\alpha} \gamma \text{ iff } \Sigma \cup \{\alpha \wedge \beta\} \vdash \gamma$$

Moreover, define $\mathcal{C}_\Sigma = \{\vdash_{\Sigma,\alpha} | \, \alpha \in Fml\}$ and the function $\mathcal{F}_\Sigma : Fml \times \mathcal{C}_\Sigma \to \mathcal{C}_\Sigma$ by $\mathcal{F}_\Sigma(\alpha, \vdash_{\Sigma,\alpha}) \vdash_{\Sigma,\alpha \wedge \beta}$. It is immediately verified, using familiar facts of classical logic such as the deduction theorem, that consequence relations as defined above satisfy all the conditions we imposed and that $\langle \mathcal{C}_\Sigma, \mathcal{F}_\Sigma \rangle$ is a CRS. We have

$$\vdash_{\Sigma,\alpha} = \vdash_{\Sigma,\beta} \text{ iff } \Sigma \vdash \alpha \leftrightarrow \beta$$

THEOREM 8.16. $\mathcal{L}_\Sigma = \langle \mathcal{C}_\Sigma, \mathcal{F}_\Sigma, \to \rangle$ *is a logic. The lattice of operators* $\mathcal{O}_{\mathcal{L}_\Sigma}$ *and thus the lattice of propositions* $\mathcal{P}_{\mathcal{L}_\Sigma}$ *are Boolean algebras isomorphic to the Lindenbaum algebra* $\mathcal{B}(\Sigma)$.

Proof. For the first part of our claim we need to prove that $\to$ is an internalising connective for $\langle \mathcal{C}_{L,\Sigma}, \mathcal{F}_{L,\Sigma} \rangle$. But this is exactly what the deduction theorem says:

$$\Sigma \cup \{\alpha\} \vdash (\beta \to \gamma) \text{ iff } \Sigma \cup \{\alpha \wedge \beta\} \vdash \gamma$$

It follows from the fact that $\to$ is internalising and Theorem 8.13 that the lattices under consideration are Boolean algebras. Moreover, it is straightforward to prove that the following function $\varphi : \mathcal{O}_{\mathcal{L}_\Sigma} \to \mathcal{B}(\Sigma)$ is well defined and is an isomorphism

$$\varphi(\overline{\alpha}) = [\alpha]_\Sigma,$$

where $[\alpha]_\Sigma$ denotes the (unique) element of the Lindenbaum algebra $\mathcal{B}(\Sigma)$ to which α belongs. ∎

We have established the well known fact that the Lindenbaum algebra is a Boolean algebra in a way, however, which is radically different from the usual proof. The reader may compare this to section 1.7

1. CONSEQUENCE REVISION SYSTEMS

1.4 The semantics of consequence revision systems: $\mathcal{H}$-Models

The Concept of an $\mathcal{H}$-Model

We now propose the following concept of a model for a CRS which heavily relies on the semantic concepts which naturally arise in nonmonotonic logic, see definition 1.33.

DEFINITION 8.17. Let $\langle \mathcal{C}, F \rangle$ be a CRS and $\mathcal{H} = \langle H, h, \mathcal{F}, l, g \rangle$ a structure such that

- H is a non-empty set
- $h : H \to \mathcal{C}$ is a surjective function
- $\mathcal{F} : Fml \times H \to H$ is a function inducing F on $Fml \times \mathcal{C}$ via h, i.e. $F(\alpha, h(x)) = h(\mathcal{F}(\alpha, x))$
- l is a function assigning to every $x \in H$ a set of Scott-models.
- g is an injective function assigning to every $x \in H$ a binary relation $\leq_x \subset H \times H$ such that $\mathcal{M}_x = \langle H, \leq_x, l \rangle$ is a $GKLM$ model for $h(x)$.

Then we say that $\mathcal{H}$ is an $\mathcal{H}$-model for $\langle \mathcal{C}, F \rangle$. We say that $\mathcal{H}$ is an $\mathcal{H}$-model for the logic $\langle \mathcal{C}, F, \rightsquigarrow \rangle$ if $\mathcal{H}$ is an $\mathcal{H}$-model for $\langle \mathcal{C}, F \rangle$. For $x \in H$ and $\alpha \in Fml$ define

$$\langle \mathcal{H}, x \rangle \models \alpha \text{ iff } s(\alpha) = 1 \text{ for all } s \in l(x).$$

The following propositions serve to illustrate the nature of $\mathcal{H}$-models. The proofs are obvious from the definition of an $\mathcal{H}$-model.

PROPOSITION 8.18. *Let $\langle \mathcal{C}, F \rangle$ be a CRS and $\mathcal{H}$ be an $\mathcal{H}$-model for $\mathcal{C}, F \rangle$. Let $\mathord{\vdash} \in \mathcal{C}$ and $x \in H$ such that $h(x) = \mathord{\vdash}$. Then the following conditions are equivalent.*

(i) $\alpha \mathrel{\vdash} \beta$

(ii) $\mathcal{M}_x \models \alpha \mathrel{\vdash} \beta$

(iii) $\mathcal{M}_{\mathcal{F}(\alpha, x)} \models \mathrel{\vdash} \beta$

(iv) $(\mathcal{H}, \mathcal{F}(\alpha, x)) \models \beta$

(v) $\mathrel{\vdash}_\alpha \beta$

PROPOSITION 8.19. *Let $\mathcal{L} = \langle \mathcal{C}, F, \rightsquigarrow \rangle$ be a logic and $\mathcal{H}$ an $\mathcal{H}$-model for $\langle \mathcal{C}, F \rangle$. Let $x \in$. Then the following conditions are equivalent*

(i) $\alpha \vdash_{\mathcal{M}_x} (\beta \rightsquigarrow \gamma)$

(ii) $\langle \mathcal{H}, \mathcal{F}(\alpha, x) \rangle \models \beta \rightsquigarrow \gamma$

(iii) $\langle \mathcal{H}, x \rangle \models \alpha \rightsquigarrow (\beta \rightsquigarrow \gamma)$

(iv) $\beta \vdash_{\mathcal{M}_{\mathcal{F}(\alpha,x)}} \gamma$

(v) $\beta \mathrel{\vert\!\sim} \gamma$, where $\mathrel{\vert\!\sim} = F(\alpha, h(x))$.

(v) $\alpha \mathrel{\vert\!\sim} (\beta \rightsquigarrow \gamma)$ with $\mathrel{\vert\!\sim} = h(x)$

The Fibred Mode of Evaluation in $\mathcal{H}$-Models

The notion of an $\mathcal{H}$-model serves a double purpose. First, it makes sense to speak of the *truth* of a formula in such a model as we are used to from traditional logics and their model theory.

Second, these models reflect the following feature of our logics. An internalising connective serves to reflect the meta-concept of consequence at the object level. So, intuitively, formulas containing the internalising connective 'talk' about consequence. $\mathcal{H}$-models account for this in that they not only model the truth of such formulas but also explicitly model the statements about consequence these formulas make. This means that in the process of evaluation of a formula in an $\mathcal{H}$-model the internalising connective is evaluated in a $GKLM$ model.

Given an $\mathcal{H}$-model $\mathcal{H}$, $x \in H$ and a formula of the form $\alpha \rightsquigarrow \beta$. We then have two ways of evaluating the internalising connective $\rightsquigarrow$. The first way of doing this is to check whether $\langle \mathcal{H}, x \rangle \models (\alpha \rightsquigarrow \beta)$ according to the definition of truth given above. The second way of evaluating the connective $\rightsquigarrow$ is to look at the $GKLM$ model $\mathcal{M}_x$ and check whether $\alpha \vdash_{\mathcal{M}_x} \beta$. If so, we have, since $\mathcal{M}_x$ is a $GKLM$ model for $h(x) =: \mathrel{\vert\!\sim}$, $\alpha \mathrel{\vert\!\sim} \beta$. We have $\alpha \vdash_{\mathcal{M}_x} \beta$ iff $\langle \mathcal{H}, x \rangle \models (\alpha \rightsquigarrow \beta)$. This is how the $\mathcal{H}$-model reflects the fact that $\rightsquigarrow$ is an internalising connective.

Let us now look at a more complex formula. Consider a formula of the form

$$\varphi = (\alpha \rightsquigarrow (\beta \rightsquigarrow (\gamma \rightsquigarrow \delta)))$$

We may now proceed as follows.
We evaluate φ in the $GKLM$ model $\mathcal{M}_x$. We have

$$\alpha \vdash_{\mathcal{M}_x} (\beta \rightsquigarrow (\gamma \rightsquigarrow \delta))$$

$$\text{iff } \beta \vdash_{\mathcal{M}_{\mathcal{F}(\alpha,x)}} (\gamma \rightsquigarrow \delta)$$

$$\text{iff } \gamma \vdash_{\mathcal{M}_{\mathcal{F}(\beta,(\mathcal{F}(\alpha,x)))}} \delta$$

1. CONSEQUENCE REVISION SYSTEMS

We have

$$\langle \mathcal{H}, x \rangle \models \varphi \text{ iff } \gamma \vdash_{\mathcal{M}_{\mathcal{F}(\beta, \mathcal{F}(\alpha, x))}} \delta$$

The characteristic feature of the second mode of evaluation is that the connective $\rightsquigarrow$ is evaluated in $GKLM$ models as consequence. At each stage in the process of evaluation we have to switch from one $GKLM$ model to another using the 'fibring function'

$$\mathcal{F}^* : Fml \times \mathcal{M} \to \mathcal{M}, \text{ where } \mathcal{M} =: \{\mathcal{M}_x \mid x \in H\}$$

defined by

$$\mathcal{F}^*(\alpha, \mathcal{M}_x) = \mathcal{M}_{\mathcal{F}(\alpha, x)}$$

Note that this 'fibring function' is well defined since by the last clause of the definition of an $\mathcal{H}$-model we have $\mathcal{M}_x = \mathcal{M}_{x'}$ iff $x = x'$. The mode of evaluation just presented is in the spirit of what was put forward by Gabbay in several papers and his books [21] and [22] as fibred semantics.

1.5 $\mathcal{H}$-models in classical logic

We will now show that the concept of an $\mathcal{H}$-model arises in a natural way in connection with the logics $\mathcal{L}_\Sigma = \langle \mathcal{C}_\Sigma, \mathcal{F}_\Sigma, \rightarrow \rangle$, i.e. classical logic. In chapter 9 we will see how this concept occurs naturally in connection with the logics arising from Hilbert spaces.

DEFINITION 8.20. Let Σ be a set of formulas consistent in classical logic. Consider the structure $\mathcal{H}_\Sigma =: \langle \mathcal{C}_\Sigma, h, \mathcal{F}_\Sigma, l_\Psi, g \rangle$ such that

- h is the identity function.

- The function l is defined as follow. $l(\vdash_{\Sigma, \alpha}) = \{s_\alpha\}$, where $s_\alpha(\beta) = 1$ iff $\vdash_{\Sigma, \alpha} \beta$, else 0.

- The function g is defined as follows. Given $x = \vdash_{\Sigma, \alpha} \in \mathcal{C}_\Sigma$, then define $g(x) =: \leq_\alpha$ as follows: $\vdash_{\Sigma, \beta} \leq_\alpha \vdash_{\Sigma, \gamma}$ is defined only if $\vdash_{\Sigma, \beta} \alpha$. Then, if $\vdash_{\Sigma, \gamma} \alpha$, $\vdash_{\Sigma, \beta} \leq_\alpha \vdash_{\Sigma, \gamma}$ iff $\vdash_{\Sigma, \gamma} \beta$. If not $\vdash_{\Sigma, \gamma} \alpha$, then $\vdash_{\Sigma, \beta} \leq_\alpha \vdash_{\Sigma, \gamma}$

Note that in the above definition the function l and g are well defined. This is readily seen using familiar facts of classical logic.

Note that we use the notation $[\alpha]$ in two different contexts, namely in the context of a CRS and in the context of a $GKLM$ model. In the present situation the notions coincide, since we have $\vdash_{\Sigma, \alpha} \beta$ iff $s_\alpha(\beta) = 1$.

LEMMA 8.21. *If $\vdash_{\Sigma, \beta} \leq_\alpha \vdash_{\Sigma, \gamma}$ and $\vdash_{\Sigma, \gamma} \leq_\alpha \vdash_{\Sigma, \beta}$, then $\vdash_{\Sigma, \beta} = \vdash_{\Sigma, \gamma}$.*

Proof. Assume that $\vdash_{\Sigma,\beta} \leq_\alpha \vdash_{\Sigma,\gamma}$ and $\vdash_{\Sigma,\gamma} \leq_\alpha \vdash_{\Sigma,\beta}$. Then we observe, inspecting the definition of $\leq_\alpha$, that we have both $\vdash_{\Sigma,\beta} \alpha$ and $\vdash_{\Sigma,\gamma} \alpha$. But in this case, again by the definition of $\leq_\alpha$, the above is only possible if $\vdash_{\Sigma,\beta} \gamma$ and $\vdash_{\Sigma,\gamma} \beta$. From this it follows by classical logic that $\vdash_{\Sigma,\beta} = \vdash_{\Sigma,\gamma}$. ■

LEMMA 8.22. $\vdash_{\Sigma,\alpha \wedge \beta}$ *is the unique* $\leq_\alpha$*-minimal element in* $[\beta]$, *where* $[\beta]$ *denotes the proposition represented by* β *in the logic* $\mathcal{L}_\Sigma$.

Proof. First note that $\vdash_{\Sigma,\alpha \wedge \beta} \in [\beta]$, since $\vdash_{\Sigma,\alpha \wedge \beta} \beta$. Clearly, $\vdash_{\Sigma,\alpha \wedge \beta} \alpha$. Let $\vdash_{\Sigma,\gamma} \in [\beta]$. If not $\vdash_{\Sigma,\gamma} \alpha$ then $\vdash_{\Sigma,\alpha \wedge \beta} \leq_\alpha \vdash_{\Sigma,\gamma}$. If $\vdash_{\Sigma,\gamma} \alpha$ then, since $\vdash_{\Sigma,\gamma} \beta$, $\vdash_{\Sigma,\gamma} \alpha \wedge \beta$. But this means $\vdash_{\Sigma,\alpha \wedge \beta} \leq_\alpha \vdash_{\Sigma,\gamma}$ for every $\vdash_{\Sigma,\gamma} \in [\beta]$. From this and the last lemma it follows that $\vdash_{\Sigma,\alpha \wedge \beta}$ a $\leq_\alpha$-minimal element in $[\beta]$. To see that it is unique, let $\vdash_{\Sigma,\delta}$ be any $\leq_\alpha$-minimal element of $[\beta]$. We have $\vdash_{\Sigma,\alpha \wedge \beta} \leq_\alpha \vdash_{\Sigma,\delta}$. Since $\vdash_{\Sigma,\delta}$ is $\leq_\alpha$-minimal in $[\beta]$ we get $\vdash_{\Sigma,\delta} = \vdash_{\Sigma,\alpha \wedge \beta}$. ■

THEOREM 8.23. $\mathcal{H}_\Sigma$ *is an* $\mathcal{H}$*-model for* $\mathcal{L}_\Sigma$.

Proof. We need to prove that for $x := \vdash_{\Sigma,\alpha} \in \mathcal{C}_\Sigma$, $\mathcal{M}_x = \langle \mathcal{C}_\Sigma, \leq_x, l \rangle$ is a $GKLM$ model for x. We have smoothness by lemma 8.22

Suppose $\beta \vdash_{\Sigma,\alpha} \gamma$. By definition this means $\Sigma \cup \{\alpha \wedge \beta\} \vdash \gamma$, which is equivalent to $\vdash_{\Sigma,\alpha \wedge \beta} \in [\gamma]$ and the claim follows from lemma 8.22 and definition 1.33. ■

2 The concept of a holistic logic

Let us start from our motivating example. In that example the consequence relations cannot be 'characterised' by a single formula, i.e. given any $\vdash_\alpha$, then there exists no formula β such that β is provable in $\vdash_\alpha$ and only in $\vdash_\alpha$. We take this observation as a motivation for studying logics in which every consequence relation has a 'characterising' formula.

DEFINITION 8.24. Given a logic $\mathcal{L} = \langle \mathcal{C}, F, \rightsquigarrow \rangle$. Consider the following conditions.

- For any non-zero $\mathord{\vdash}_0 \in \mathcal{C}$ there exists a formula $\sigma_{\mathord{\vdash}_0}$ such that $\mathord{\vdash} \sigma_{\mathord{\vdash}_0}$ iff $\mathord{\vdash} = \mathord{\vdash}_0$. We call σ a pointer to $\mathord{\vdash}$.

- For every $\mathord{\vdash} \in \mathcal{C}$ there exist a formula α such that neither $\mathord{\vdash} \alpha$ nor $\mathord{\vdash} \neg \alpha$.

We call $\mathcal{L}$ a (non-degenerate) holistic logic if both of the above conditions are satisfied. We call $\mathcal{L}$ degenerate if the first condition is satisfied but not the second. We call $\mathcal{L}$ totally degenerate if the first condition is satisfied but for no consequence relation $\mathord{\vdash}$ does there exist an α such that neither α nor $\mathord{\vdash} \neg \alpha$.

2. THE CONCEPT OF A HOLISTIC LOGIC

Remarks: Any two pointers σ_1 and σ_2 to the same consequence relation are equivalent, i.e. $[\sigma_1] = [\sigma_2]$ We assume the consequence relation referred to later to be non-zero, i.e. consistent without explicit mentioning.

Intuitively, the second condition says that every consequence relation $\mathrel{\mid\!\sim}$ must be genuinely revisable, i.e. we assume that there exists a formula α such that $\mathrel{\mid\!\sim}_\alpha$ is consistent and distinct from $\mathrel{\mid\!\sim}$. It follows that a (non-degenerate) holistic logic has at least two consequence relations. We will always use the term 'holistic' in the sense of 'non-degenerate holistic' except in the theorem which we call the limiting case theorem. In the case of a totally degenerate holistic logic there is no genuine revision at all.

In the next subsections we state some salient properties of holistic logics.

2.1 Orthogonality, encodedness, dimension

DEFINITION 8.25. Let $\mathcal{L}$ be a holistic logic and let $\mathrel{\mid\!\sim}_1$ and $\mathrel{\mid\!\sim}_2$ be two consequence relations of $\mathcal{L}$ with pointers σ_1 and σ_2 respectively. Then we say that $\mathrel{\mid\!\sim}_1$ and $\mathrel{\mid\!\sim}_2$ are orthogonal if $\mathrel{\mid\!\sim}_1 \neg\sigma_2$ and $\mathrel{\mid\!\sim}_2 \neg\sigma_1$.

Actually it suffices to require one of the two conditions. It can then be proved using the second of the global conditions we impose on the consequence relations that the relation of orthogonality is symmetric.

The following lemma follows from the definition of a pointer and that of a consequence revision systems.

LEMMA 8.26. *Let* $\mathcal{L} = \langle \mathcal{C}, F, \rightsquigarrow \rangle$ *be a holistic logic,* $\mathrel{\mid\!\sim}_1, \mathrel{\mid\!\sim}_2 \in \mathcal{C}$. *If* $\mathrel{\mid\!\sim}_1$ *and* $\mathrel{\mid\!\sim}_2$ *are not orthogonal, then* $\mathrel{\mid\!\sim}_{1,\sigma_2} = \mathrel{\mid\!\sim}_2$ *and vice versa. If they are orthogonal we have* $\mathrel{\mid\!\sim}_{1,\sigma_2} = 0$ *and vice versa.*

DEFINITION 8.27. Let $\mathcal{L}$ be a holistic logic. We call a family $(\mathrel{\mid\!\sim}_i)_{i\in I}$ of pairwise orthogonal consequence relations a basis of $\mathcal{L}$ if for any $\mathrel{\mid\!\sim}$ of $\mathcal{L}$ there exists a basis consequence relation $\mathrel{\mid\!\sim}_j$ not orthogonal to $\mathrel{\mid\!\sim}$. We call $\mathcal{L}$ finite-dimensional iff it admits a finite basis. If $\mathcal{L}$ is finite-dimensional we say it has dimension n if it admits a basis of n elements and no basis of fewer elements.

Remark: We may view the basis consequence relations $\mathrel{\mid\!\sim}_i$ as containing all the 'information' of the logic $\mathcal{L}$ in the sense that every consequence relation is encoded in at least one basis consequence relation (see theorem 8.30).

LEMMA 8.28. *Let* $\mathcal{L} = \langle \mathcal{C}, F, \rightsquigarrow \rangle$ *be a holistic logic and* $\mathrel{\mid\!\sim} \in \mathcal{C}$ *with pointer* σ. *Then we have*

- (i) $\mathrel{\mid\!\sim} \alpha$ *iff* $[\sigma \rightsquigarrow \alpha] = [\top]$ *and thus* $[\neg(\sigma \rightsquigarrow \alpha)] = [\bot]$

- (ii) $\mathrel{\mid\!\not\sim} \alpha$ *iff* $[\sigma \rightsquigarrow \alpha] = [\neg\sigma]$ *and thus* $[\neg(\sigma \rightsquigarrow \alpha)] = [\sigma]$

Remark: Note that by the above lemma we have that $\mathrel{\vdash\mkern-9mu\sim} \alpha$ iff $\mathrel{\vdash\mkern-9mu\sim} \sigma \rightsquigarrow \alpha$ and $\mathrel{\not\vdash\mkern-9mu\sim} \alpha$ iff $\mathrel{\vdash\mkern-9mu\sim} \neg(\sigma \rightsquigarrow \alpha)$. We may therefore view the formula $\sigma \rightsquigarrow \alpha$ as expressing provability of α at the object level and the formula $\neg(\sigma \rightsquigarrow \alpha)$ as expressing the unprovability of α at the object level. In particular we have 'provability of unprovability' in the sense that if $\mathrel{\vdash\mkern-9mu\sim}$ cannot prove α, then it can prove that it cannot prove α.

Proof. (*i*) For the direction from left to right suppose $\mathrel{\vdash\mkern-9mu\sim} \alpha$ and note that for any $\mathrel{\vdash\mkern-9mu\sim}_1$ orthogonal to $\mathrel{\vdash\mkern-9mu\sim}$ we have $\mathrel{\vdash\mkern-9mu\sim}_1 \sigma \rightsquigarrow \alpha$, see lemma 8.26. If $\mathrel{\vdash\mkern-9mu\sim}_1$ is non-orthogonal to $\mathrel{\vdash\mkern-9mu\sim}$, we have by lemma 8.26 that $\mathrel{\vdash\mkern-9mu\sim} \sigma \rightsquigarrow \alpha$. Thus $[\sigma \rightsquigarrow \alpha] = \mathcal{C} = [\top]$. The other direction is obvious.

(*ii*) Suppose that $\mathrel{\not\vdash\mkern-9mu\sim} \alpha$. Then, again, we have for every $\mathrel{\vdash\mkern-9mu\sim}_1$ orthogonal to $\mathrel{\vdash\mkern-9mu\sim}$ that $\mathrel{\vdash\mkern-9mu\sim}_1 \sigma \rightsquigarrow \alpha$. But if $\mathrel{\vdash\mkern-9mu\sim}_1$ is not orthogonal to $\mathrel{\vdash\mkern-9mu\sim}$, $\mathrel{\vdash\mkern-9mu\sim}_1 \sigma \rightsquigarrow \alpha$ cannot hold, since this would imply $\mathrel{\vdash\mkern-9mu\sim} \alpha$ contrary to the hypothesis. Thus $[\sigma \rightsquigarrow \alpha] = [\neg\sigma]$. The other direction is obvious. ∎

Remark: Note that the propositions $[\sigma], [\neg\sigma], [\top], [\bot]$ do not depend on the pointer σ, since any two pointers are equivalent. They form a Boolean algebra in a natural way.

PROPOSITION 8.29. *Assume the hypotheses of the last lemma and let φ and ψ have the form $\varphi = \sigma \rightsquigarrow \alpha$ or $\varphi = \neg(\sigma \rightsquigarrow \alpha)$ and $\psi = \sigma \rightsquigarrow \beta$ or $\psi = \neg(\sigma \rightsquigarrow \beta)$. Then we have*

- (*i*) $\mathrel{\vdash\mkern-9mu\sim} \neg\varphi$ *iff* $\mathrel{\not\vdash\mkern-9mu\sim} \varphi$

- (*ii*) $\mathrel{\vdash\mkern-9mu\sim} \varphi \wedge \psi$ *iff* $\mathrel{\vdash\mkern-9mu\sim} \varphi$ *and* $\mathrel{\vdash\mkern-9mu\sim} \psi$

- (*iii*) $\mathrel{\vdash\mkern-9mu\sim} \varphi \vee \psi$ *iff* $\mathrel{\vdash\mkern-9mu\sim} \varphi$ *or* $\mathrel{\vdash\mkern-9mu\sim} \psi$

- (*iv*) $\mathrel{\vdash\mkern-9mu\sim} \varphi \rightarrow \psi$ *iff* $\mathrel{\not\vdash\mkern-9mu\sim} \varphi$ *or* $\mathrel{\vdash\mkern-9mu\sim} \psi$

- (*v*) $\varphi \mathrel{\vdash\mkern-9mu\sim} \psi$ *iff* $\mathrel{\vdash\mkern-9mu\sim} \varphi \rightarrow \psi$

- (*vi*) $\mathrel{\vdash\mkern-9mu\sim} \varphi \rightarrow \psi$ *iff* $\mathrel{\vdash\mkern-9mu\sim} \varphi \rightsquigarrow \psi$

Proof. For (*i*) we reason as follows. Let φ have the form $\varphi = \sigma \rightsquigarrow \alpha$ and suppose $\mathrel{\vdash\mkern-9mu\sim} \neg\varphi$. Then, clearly, $\mathrel{\not\vdash\mkern-9mu\sim} \varphi$. Now suppose $\mathrel{\not\vdash\mkern-9mu\sim} \varphi$. By lemma 8.28 we then have $[\neg\varphi] = [\sigma]$. It follows that $\mathrel{\vdash\mkern-9mu\sim} \neg\varphi$.

(*ii*) is a general property of the consequence relations considered. (*iii*) and (*iv*) follow from (*i*) and the definition of the connectives $\vee$ and $\rightarrow$. In order to see that (*v*) holds recall that $\varphi \mathrel{\vdash\mkern-9mu\sim} \psi$ means that $\mathrel{\vdash\mkern-9mu\sim}_\varphi \psi$ and observe that by lemma 8.28 $\mathrel{\vdash\mkern-9mu\sim}_\varphi = \mathrel{\vdash\mkern-9mu\sim}$, namely if $\mathrel{\vdash\mkern-9mu\sim} \varphi$, or $\mathrel{\vdash\mkern-9mu\sim}_\varphi = 0$, namely if $\mathrel{\not\vdash\mkern-9mu\sim} \varphi$. (*vi*) follows from (*v*) and the fact that $\rightsquigarrow$ is an internalising connective. ∎

2. THE CONCEPT OF A HOLISTIC LOGIC

THEOREM 8.30. *Let $\mathcal{L}$ be a holistic logic and $\mathrel{\mid\!\sim}_1$ and $\mathrel{\mid\!\sim}_2$ two non-orthogonal consequence relations with pointers σ_1 and σ_2 respectively. Then we have*

- (i) $\alpha \mathrel{\mid\!\sim}_1 \beta$ iff $\mathrel{\mid\!\sim}_2 \sigma_1 \rightsquigarrow (\alpha \rightsquigarrow \beta)$
- (ii) $\alpha \mathrel{\mid\!\not\sim}_1 \beta$ iff $\mathrel{\mid\!\sim}_2 \sigma_1 \rightsquigarrow (\neg(\sigma_1 \rightsquigarrow (\alpha \rightsquigarrow \beta)))$

By symmetry the claim also holds if we interchange the indices 1 and 2.

Proof. (i) Recall that $\mathrel{\mid\!\sim}_1$ and $\mathrel{\mid\!\sim}_2$ are non-orthogonal iff $\mathrel{\mid\!\sim}_{2_{\sigma_1}} = \mathrel{\mid\!\sim}_1$ (and vice versa). We have $\alpha \mathrel{\mid\!\sim}_2 \beta$ iff $\mathrel{\mid\!\sim}_2 \alpha \rightsquigarrow \beta$, since $\rightsquigarrow$ is internalising. $\alpha \mathrel{\mid\!\sim}_2 \beta$ is thus equivalent to $\mathrel{\mid\!\sim}_{2_{\sigma_1}} \alpha \rightsquigarrow \beta$. This is the case iff $\sigma_1 \mathrel{\mid\!\sim}_2 \alpha \rightsquigarrow \beta$, which is equivalent to $\mathrel{\mid\!\sim}_2 \sigma_1 \rightsquigarrow (\alpha \rightsquigarrow \beta)$.

(ii) Note that $\alpha \mathrel{\mid\!\not\sim}_1 \beta$ is by 'provability of unprovability' equivalent to $\mathrel{\mid\!\sim}_1 \neg(\sigma_1 \rightsquigarrow (\alpha \rightsquigarrow \beta))$ and apply (i). ∎

The above theorem says that non-orthogonal consequence relations of a holistic logic are 'encoded' in each other. This fact is the motivation for calling these structures holistic.

2.2 Self-referential soundness and completeness

In this section we study self-referential soundness and completeness in holistic logics. This notion was, essentially, first introduced by in [57] and [58] for modal systems. We will prove that the consequence relations of a holistic logic are self-referentially sound and complete and, with a certain natural exception, nonmonotonic.

We now define a meta language in which we can talk about provability. Intuitively, $DER(\alpha, \beta)$ means "β is derivable from α in $\mathrel{\mid\!\sim}$".

DEFINITION 8.31.

- (i) If α, β are formulas of the object language, then $DER(\alpha, \beta) \in ML$.
- If α is a formula of the object language and $\varphi \in ML$, then $DER(\alpha, \varphi) \in ML$ and $DER(\varphi, \alpha) \in ML$.
- If $\varphi, \psi \in ML$, then $DER(\varphi, \psi) \in ML$.
- If $\varphi, \psi \in ML$, so are $\neg\varphi$ and $\varphi \wedge \psi$, $\varphi \vee \psi$, $\varphi \rightarrow \psi$, where $\vee$ and $\rightarrow$ are defined as usual in terms of $\neg$ and $\wedge$.

We will use the following abbreviations:

$$PROV\alpha =: DER(\top, \alpha)$$

$$CON\alpha =: \neg PROV \neg \alpha$$

$$EQUIV(\alpha, \beta) =: DER(\alpha, \beta) \wedge DER(\beta, \alpha)$$

We now define a natural translation of the meta language ML into the object language. We assume that we have a logic $\mathcal{L} = \langle \mathcal{C}, F, \rightsquigarrow \rangle$. The following definitions are relative to a fixed $\mathrel{|\!\!\sim} \in \mathcal{C}$ having a pointer σ to itself. Since any two pointers are equivalent, they do not 'depend' on the pointer chosen.

DEFINITION 8.32. Let σ be a pointer to $\mathrel{|\!\!\sim}$. Define the translation $'$ as follows.

- (i) If $\varphi = DER(\alpha, \beta)$ where α and β are formulas of the object language, $\varphi' =: \sigma \rightsquigarrow (\alpha \rightsquigarrow \beta)$.

- (ii) If $\varphi = DER(\alpha, \psi)$, where α is a formula of the object language and $\psi \in ML$, then $\varphi' =: \sigma \rightsquigarrow (\alpha \rightsquigarrow \psi')$; analogously for the case $DER(\psi, \alpha)$.

- (iii) If $\varphi = DER(\psi, \rho)$ with $\psi, \rho \in ML$, $\varphi' =: \sigma \rightsquigarrow (\psi' \rightsquigarrow \rho')$.

- (iv) If $\varphi = \neg\psi$, $\varphi' =: \neg(\sigma \rightsquigarrow \psi')$.

- (v) If $\varphi = \psi \wedge \rho$, $\varphi' =: \psi' \wedge \rho'$.

We now define the notion of truth for ML in a natural way. This definition of truth is in the spirit of what Smullyan calls a self-referential interpretation in the above mentioned books. The essential feature of Smullyan's notion of self-referential truth is this. Given a modal system M with the modal operator $\Box$. Then we say that a formula of the form $\Box A$ is (self-referentially) true with respect to M iff A is provable in M.

DEFINITION 8.33.

- (i) If $\varphi = DER(\alpha, \beta)$, where α, β are formulas of the object language, then TRUE φ iff $\alpha \mathrel{|\!\!\sim} \beta$.

- (ii) If $\varphi = DER(\alpha, \psi)$, where α is a wff of the object language, then TRUE φ iff $\alpha \mathrel{|\!\!\sim} \psi'$; analogously for the case $DER(\psi, \alpha)$.

- (iii) If $\varphi = DER(\psi, \rho)$ for $\psi, \rho \in ML$, then TRUE φ iff $\psi' \mathrel{|\!\!\sim} \rho'$.

- (iv) If $\varphi = \neg\psi$, then TRUE φ iff not TRUE ψ.

- (v) If $\varphi = \psi \wedge \rho$, then TRUE φ iff TRUE ψ and TRUE ρ.

2. THE CONCEPT OF A HOLISTIC LOGIC

THEOREM 8.34. *Let* $\mathcal{L} = \langle \mathcal{C}, F, \leadsto \rangle$ *be a holistic logic,* $\mathrel{\mid\!\sim} \in \mathcal{C}$. *Then we have for any* $\mathrel{\mid\!\sim} \in \mathcal{C}$ *and any* $\varphi \in ML$

$$TRUE\ \varphi\ \text{iff}\ \mathrel{\mid\!\sim} \varphi'$$

The above theorem expresses *self-referential soundness and completeness* of the consequence relations of a holistic logic. The fact that $\mathrel{\mid\!\sim} \varphi'$ implies $TRUE\ \varphi$ expresses and the fact that $TRUE\ \varphi$ implies $\mathrel{\mid\!\sim} \varphi'$ expresses .

Proof. By induction on the construction of the formulas of ML.
(i) Case $\varphi = DER(\alpha, \beta)$. By definition $TRUE\ \varphi$ means $\alpha \mathrel{\mid\!\sim} \beta$. This means $\mathrel{\mid\!\sim} \alpha \leadsto \beta$, which is equivalent to $\mathrel{\mid\!\sim} \sigma \leadsto (\alpha \leadsto \beta)$. But this says that $\mathrel{\mid\!\sim} \varphi'$.
(ii) Case $\varphi = DER(\alpha, \psi)$. Suppose $TRUE\ \varphi$. By definition this says $\alpha \mathrel{\mid\!\sim} \psi'$ or equivalently $\mathrel{\mid\!\sim} \sigma \leadsto (\alpha \leadsto \psi')$. But this is exactly what $\mathrel{\mid\!\sim} \varphi'$ means.
(iii) The case $\varphi = DER(\psi, \rho)$ is proved analogously to that of (ii).
(iv) Case $\varphi = \neg\psi$. $TRUE\ \varphi$ means that not $TRUE\ \psi$. By the induction hypothesis this is equivalent to $\mathrel{\mid\!\not\sim} \psi'$. This is by lemma 8.28 the case iff $\mathrel{\mid\!\sim} \neg(\sigma \leadsto \psi')$. But this says $\mathrel{\mid\!\sim} \varphi'$.
(v) Case $\varphi\psi \wedge \rho$. We have by definition $TRUE\ \varphi$ iff $TRUE\ \psi$ and $TRUE\ \rho$. The latter is by the induction hypothesis equivalent to "$\mathrel{\mid\!\sim} \varphi'$ and $\mathrel{\mid\!\sim} \rho'$" which in turn is equivalent to $\mathrel{\mid\!\sim} \varphi'$. ∎

Remark: Inspecting the translation of the metalanguage into the object language, we may view the metalanguage as a 'sublanguage' of the object language. The peculiar feature of this 'sublanguage' is that it contains a 'proof operator' □, namely $\Box\alpha :=\ \sigma \leadsto \alpha$, as opposed to 'proof predicates' which we have in other languages.

Instead of explicitly defining a metalanguage we could have proceeded as follows. We could from the outset have confined ourselves to defining a sublanguage MSL of the object language doing the job of the metalanguage. In this case the expression $DER(\alpha, \beta)$ would be an *abbreviation* for $\sigma \leadsto (\alpha \leadsto \beta)$. We could then have defined $TRUE\ \varphi$ directly for the formulas of MSL, i.e. for object formulas. Theorem 8.34 would then read: $TRUE\ \varphi$ iff $\mathrel{\mid\!\sim} \varphi$.

The notion of self-referentiality thus becomes fully analogous to that introduced by Smullyan in connection with self-application of modal systems, where the modal operator □ plays the role of a proof operator.

Example: Let us consider an example and let us for the sake of illustration verify the truth of the claim made in the above theorem directly. Let α be an object formula and consider the following meta-statement

$$\varphi = PROV\alpha \to CON\alpha$$

Its translation is

$$\varphi' = (\sigma \leadsto (\top \leadsto \alpha)) \to \neg(\sigma \leadsto (\sigma \leadsto (\top \leadsto \neg\alpha)))$$

Let us first verify that $TRUE\ \varphi$ implies $\mathrel{\mid\!\sim} \varphi'$. Assume that not $TRUE\ PROV\alpha$. This means that $TRUE\neg PROV\alpha$, which says that $\mathrel{\mid\!\not\sim} \alpha$. By Lemma 8.28 we have $[\neg(\sigma \leadsto (\top \leadsto \alpha))] = [\sigma]$. Thus $[\varphi'] = [\sigma \vee ...]$ and we have $\mathrel{\mid\!\sim} \varphi'$.
Now assume $TRUE\ CON\alpha$, i.e. $\mathrel{\mid\!\not\sim} \neg\alpha$ and thus $\mathrel{\mid\!\not\sim} \top \leadsto \neg\alpha$, hence $\mathrel{\mid\!\not\sim} \sigma \leadsto (\top \leadsto \neg\alpha)$. In this case we have by lemma 8.28 $[\neg(\sigma \leadsto (\sigma \leadsto (\top \leadsto \neg\alpha)))] = [\sigma]$. Thus $[\varphi'] = [... \vee \sigma]$ and we have $\mathrel{\mid\!\sim} \varphi'$.
Let us now verify that $\mathrel{\mid\!\sim} \varphi'$ implies $TRUE\ \varphi$. So assume $\mathrel{\mid\!\sim} \varphi'$. $[\neg(\sigma \leadsto (\sigma \leadsto (\top \leadsto \alpha)))]$ equals either $[\bot]$ or $[\sigma]$. In the first case we have $\mathrel{\mid\!\sim} \alpha$. Since $\mathrel{\mid\!\sim}$ is assumed to be consistent, we have $\mathrel{\mid\!\not\sim} \neg\alpha$, which means $TRUE\ CON\alpha$. But this says that $TRUE\ \varphi$.
In the second case we have $\mathrel{\mid\!\not\sim} \alpha$ and thus not $TRUE\ PROV\alpha$ in which case again $TRUE\ \varphi$.

Some examples of true meta-statements

PROPOSITION 8.35.

The following meta-statements are true and thus their translations are provable.

- $\varphi_1 = PROV\varphi \leftrightarrow PROV\,PROV\varphi$

- $\varphi_2 = \neg PROV\varphi \leftrightarrow PROV\neg PROV\varphi$

- $\varphi_3 = CON\varphi \to \neg EQUIV(\varphi, \neg PROV\varphi)$

- $\varphi_4 = (CON\varphi \wedge \neg PROV\varphi) \to (PROV\neg PROV\varphi \wedge \neg DER(\varphi, \neg PROV\varphi)$

- $\varphi_5 = PROV\varphi \leftrightarrow EQUIV(\varphi, \neg PROV\bot)$

- $\varphi_6 = PROV\neg PROV\bot$

- $\varphi_6 = (PROV\alpha \wedge DER(\varphi, \psi)) \to PROV\psi$

Comment: The above claims are immediate consequences of lemma 8.28. The reader should note that by self-referential completeness the consequence relation 'knows' the facts expressed by the above statements.
Intuitively, φ_1 expresses 'provability of provability': φ is provable iff it is provable that φ is provable.
φ_2 expresses 'provability of unprovability': φ is not provable iff its unprovability can be proved.

φ_3 says that if φ is consistent, then it is not equivalent to its 'own unprovability'. This says that the consequence relations of a holistic logic do not admit Gödel fixed points.

φ_4 says the following. Suppose φ is consistent and not provable. Then we know that its unprovability can be proved. What φ_4 says is that, however, its unprovability cannot be proved 'from φ'. So, φ_4 can be rephrased as follows. If φ is consistent and not provable, then its unprovability can be proved but not from φ. In the non-degenerate case φ_4 says in particular that the consequence relation is nonmonotonic, since in the non-degenerate case we assume it to have an object formula which is consistent and not provable.

φ_5 says that φ is provable iff it is equivalent to the consistency of the consequence relation.

φ_6 says that the consequence relation can prove its consistency.

φ_7 expresses modus ponens for meta-statements. In a sense, the consequence relation can 'justify' the logical rule of modus ponens. Normally, logical rules such as modus ponens are justified at the meta level as preserving truth. The intuitive meaning of φ_7 is that holistic logics can *prove* its own rules (at the object level).

The case of a complete classical theory

Recall the definition of $\mathcal{L}_\Sigma = \langle \mathcal{C}_\Sigma, \mathcal{F}_\Sigma, \rightarrow \rangle$ from the motivating example. We have the

PROPOSITION 8.36. *Let Σ be a consistent set of formulas. Then $\mathcal{L}_\Sigma = \langle \mathcal{C}_\Sigma, \mathcal{F}_\Sigma, \rightarrow \rangle$ is holistic iff Σ is a complete classical theory. In this case $\mathcal{L}_\Sigma$ is (totally) degenerate. It has dimension 1 and we have $\mathcal{C} = \{\vdash_\Sigma, 0\}$ and $\mathcal{F}_\Sigma(\alpha, \vdash_\Sigma) = \vdash_\Sigma$ if $\alpha \in \Sigma$, else 0.*

Proof. Observe that for any α such that neither $\Sigma \vdash \neg \alpha$ nor $\Sigma \vdash \alpha$, $\vdash_\Sigma$ is a proper subset of $\vdash_\Sigma, \alpha$. So in this case $\vdash_\Sigma$ cannot have a pointer. It follows that $\vdash_\Sigma$ can have a pointer only if for every α either $\Sigma \vdash \alpha$ or $\Sigma \vdash \neg \alpha$, i.e. Σ is a complete theory. In fact, in this case any formula α such that $\Sigma \vdash \alpha$ is a pointer to $\vdash_\Sigma$ ∎

3 No windows theorems: second version

3.1 The local no windows theorem

Given a holistic logic $\mathcal{L} = \langle \mathcal{C}, F, \rightsquigarrow \rangle$ and $\mathord{\mid\!\sim} \in \mathcal{C}$ with pointer σ. Then we define $\Sigma_{\mid\!\sim} =: \{\alpha \mid \mathord{\mid\!\sim} \alpha\}$, to be the *local theory* of $\mathord{\mid\!\sim}$. We denote by Σ_g its *global theory*, i.e. $\Sigma_g =: \{\alpha \mid \mathord{\mid\!\sim} \alpha$ for all $\mathord{\mid\!\sim} \in \mathcal{C}\}$.

LEMMA 8.37. *Let $\mathcal{L} = \langle \mathcal{C}, F, \rightsquigarrow \rangle$ be a holistic logic and $\mathord{\mid\!\sim} \in \mathcal{C}$ with pointer σ. Suppose $\mathord{\mid\!\sim} \alpha$. Then $\sigma \rightsquigarrow \alpha \in \Sigma_g$.*

Proof. Let $\mathrel{\mid\!\sim}_1 \in \mathcal{C}$. Suppose $\mathrel{\mid\!\sim}_1$ is orthogonal to $\mathrel{\mid\!\sim}$. Then $\mathrel{\mid\!\sim}_{1\sigma} = 0$ and, clearly, $\sigma \mathrel{\mid\!\sim}_1 \alpha$. Hence $\mathrel{\mid\!\sim}_1 \sigma \rightsquigarrow \alpha$. Suppose $\mathrel{\mid\!\sim}_1$ is not orthogonal to $\mathrel{\mid\!\sim}$. In this case we have $\mathrel{\mid\!\sim}_{1\sigma} = \mathrel{\mid\!\sim}$. Thus $\mathrel{\mid\!\sim}_{1\sigma}\alpha$. It follows that $\sigma \mathrel{\mid\!\sim}_1 \alpha$ which means $\mathrel{\mid\!\sim}_1 \sigma \rightsquigarrow \alpha$. We have proved that $\sigma \rightsquigarrow \alpha \in \Sigma_g$. ∎

In the sequel we will use the terminology 'the connective $\rightsquigarrow$ is classically equivalent to material implication'. By this we simply mean that $\alpha \rightsquigarrow \beta$ is an abbreviation for a formula which is classically equivalent to $\alpha \rightarrow \beta$. For instance the Sasaki hook $\rightsquigarrow_s$ has this property because $\alpha \rightsquigarrow_s \beta =: \neg \alpha \vee (\alpha \wedge \beta)$ is classically equivalent to $\alpha \rightarrow \beta$.

LEMMA 8.38. *Let $\mathcal{L} = \langle \mathcal{C}, F, \rightsquigarrow \rangle$ be a holistic logic and $\mathrel{\mid\!\sim} \in \mathcal{C}$ with pointer σ. Suppose $\rightsquigarrow$ is classically equivalent to $\rightarrow$, i.e. material implication. Assume that $\Sigma_g \cup \{\sigma\}$ is classically consistent. Then we have $\mathrel{\mid\!\sim} \alpha$ iff $\Sigma_g \cup \{\sigma\} \vdash \alpha$.*

Proof. For the direction from left to right note that $\mathrel{\mid\!\sim} \alpha$ implies by lemma 8.37 that $\sigma \rightsquigarrow \alpha \in \Sigma_g$ and, since $\rightsquigarrow$ is assumed to be classically equivalent to $\rightarrow$, we have $\Sigma_g \cup \{\sigma\} \vdash \alpha$.

For the other direction suppose $\Sigma_g \cup \{\sigma\} \vdash \alpha$ and assume $\mathrel{\mid\!\not\sim} \alpha$. We have $\Sigma_g \cup \{\sigma\} \vdash \sigma \rightarrow \alpha$. On the other hand we have by 'provability of unprovability' $\mathrel{\mid\!\sim} \neg(\sigma \rightsquigarrow \alpha)$ and thus, by the direction already proved, $\Sigma_g \cup \{\sigma\} \vdash \neg(\sigma \rightsquigarrow \alpha)$ and thus, since $\rightsquigarrow$ is classically equivalent to $\rightarrow$ $\Sigma_g \cup \{\sigma\} \vdash \neg(\sigma \rightarrow \alpha)$

$\Sigma_g \cup \{\sigma\}$ would thus be classically inconsistent contrary to the hypothesis. It follows that $\mathrel{\mid\!\sim} \alpha$. ∎

We call the following theorem the (local) no windows theorem because it is reminiscent of what Leibniz in his *Monadology* says about the monads: "The monads have no windows".

THEOREM 8.39. *Let $\mathcal{L} = \langle \mathcal{C}, F, \rightsquigarrow \rangle$ be a non-degenerate holistic logic. Suppose $\rightsquigarrow$ is classically equivalent to material implication. Let $\mathrel{\mid\!\sim} \in \mathcal{C}$ with pointer σ. Then $\Sigma_g \cup \{\sigma\}$ is classically inconsistent. Thus, $\Sigma_{\mathrel{\mid\!\sim}}$ is classically inconsistent.*

Proof. Let σ be any pointer with corresponding $\mathrel{\mid\!\sim} \in \mathcal{C}$. Assume that $\Sigma_g \cup \{\sigma\}$ is classically consistent. Let α be such that $\mathrel{\mid\!\not\sim} \neg\alpha$ and $\mathrel{\mid\!\not\sim} \alpha$. By the hypothesis of non-degeneracy such a formula exists. Then we have by 'provability of unprovability' and nonmonotonicity

(1) $\mathrel{\mid\!\sim} \neg(\sigma \rightsquigarrow \alpha)$

(2) $\alpha \mathrel{\mid\!\not\sim} \neg(\sigma \rightsquigarrow \alpha)$

We have by lemma 8.38

3. NO WINDOWS THEOREMS: SECOND VERSION

$$(3) \quad \Sigma_g \cup \{\sigma\} \vdash \neg(\sigma \leadsto \alpha)$$

and thus by classical logic

$$(4) \quad \Sigma_g \cup \{\sigma\} \vdash \alpha \to \neg(\sigma \leadsto \alpha)$$

Since $\leadsto$ is classically equivalent to $\to$, it follows that

$$(5) \quad \Sigma_g \cup \{\sigma\} \vdash \alpha \leadsto \neg(\sigma \leadsto \alpha)$$

Again, by lemma 8.38 we get

$$(6) \quad \mathrel{\vert\!\sim} \alpha \leadsto \neg(\sigma \leadsto \alpha)$$

and thus

$$(7) \quad \alpha \mathrel{\vert\!\sim} \neg(\sigma \leadsto \alpha)$$

But (7) contradicts (2). It follows that $\Sigma_g \cup \{\sigma\}$ is classically inconsistent. ∎

3.2 The global no windows theorem

We now restrict ourselves to the case of a finite-dimensional holistic logic. In this case we can sharpen the no windows theorem so as to get a Kochen-Specker type result as a special case.

LEMMA 8.40. *Let $\mathcal{L}$ be any logic and α such that $\mathrel{\vert\!\not\sim} \alpha$ for every $\mathrel{\vert\!\sim} \neq 0$. Then $\alpha \leadsto \bot \in \Sigma_g$.*

Proof. Given any $\mathrel{\vert\!\sim} \in \mathcal{C}$ and α as in the hypothesis. We claim that $\mathrel{\vert\!\sim}_\alpha = 0$. For otherwise we would have $\mathrel{\vert\!\sim}_\alpha \alpha$ with $\mathrel{\vert\!\sim}_\alpha \neq 0$ contrary to the hypothesis. Thus $\mathrel{\vert\!\sim}_\alpha \bot$, which means $\alpha \mathrel{\vert\!\sim} \bot$ and thus $\mathrel{\vert\!\sim} \alpha \leadsto \bot$. We have proved that $\alpha \leadsto \bot \in \Sigma_g$. ∎

The following theorem is a summary of previous results and, moreover, contains the strengthened version of the no windows theorem.

THEOREM 8.41. *Let $\mathcal{L} = \langle \mathcal{C}, \mathcal{F}, \leadsto \rangle$ be a non-degenerate holistic logic. Suppose that $\leadsto$ is classically equivalent to $\to$, i.e. material implication. Then we have the following*

- *(i) Every consistent $\mathrel{\vert\!\sim} \in \mathcal{C}$ is nonmonotonic.*

- *(ii) For any $\mathrel{\vert\!\sim} \in \mathcal{C}$, $\Sigma_{\mathrel{\vert\!\sim}}$ is classically inconsistent.*

- *(iii) If $\mathcal{L}$ is finite dimensional, then Σ_g is classically inconsistent. In fact, it contains a classical contradiction.*

Proof. (*i*) and (*ii*) summarise results proved earlier.

As to (*iii*) let $(\mathrel{|\!\sim}_i), i = 1, .., n$ a basis with pointers σ_i. The local no windows theorem 8.39 tells that $\Sigma \cup \{\sigma_i\}$ is classically inconsistent, $i = 1, ..., n$. This means that

$$\Sigma_g \vdash \neg\sigma_i, \, i = 1, ..., n.$$

Therefore

$$\Sigma_g \vdash \bigwedge_i \neg\sigma_i$$

For any $\mathrel{|\!\sim} \neq 0$ we have by the definition of a basis

$$\mathrel{|\!\not\sim} \bigwedge_i \neg\sigma_i$$

For otherwise $\mathrel{|\!\sim}$ would be orthogonal to all elements of the basis contrary to the definition of a basis. It follows by lemma 8.40 that

$$\bigwedge_i \neg\sigma_i \rightsquigarrow \bot \in \Sigma_g$$

Thus

$$\Sigma_g \vdash \bigwedge_i \neg\sigma_i \rightsquigarrow \bot$$

and, since $\rightsquigarrow$ is classically equivalent to $\rightarrow$,

$$\Sigma_g \vdash \bigwedge_i \neg\sigma_i \rightarrow \bot$$

It follows that

$$\Sigma_g \vdash \bot$$

Thus Σ_g is classically inconsistent. Then there exists a finite set $\{\alpha_1, ..., \alpha_n\} \subset \Sigma_g$ which is classically inconsistent. Since Σ_g is closed under conjunctions we have

$$\bigwedge_i \alpha_i \in \Sigma_g$$

But this conjunction is a classical contradiction. ■

4 Limiting case theorem: second version

In this section we prove a limiting case theorem for holistic logics. Let us start with the following observation.

LEMMA 8.42. *Given a consequence revision system $\langle \mathcal{C}, F \rangle$. Suppose that all revision operators commute. Then every $\mathrel{|\!\sim} \in \mathcal{C}$ is monotonic.*

4. LIMITING CASE THEOREM: SECOND VERSION

Proof. Assume that all operators commute and let $\mathrel{\mid\!\sim}\, \in \mathcal{C}$. Assume $\mathrel{\mid\!\sim} \beta$. This means $\mathrel{\mid\!\sim}_\beta = \mathrel{\mid\!\sim}$. Now let α be any formula. Note that $\mathrel{\mid\!\sim}_{\alpha,\beta} \beta$. The above notation means that $\mathrel{\mid\!\sim}$ is first revised by α and then by β. Since the revision operators corresponding to α and β commute, we have $\mathrel{\mid\!\sim}_{\beta,\alpha} = \mathrel{\mid\!\sim}_{\alpha,\beta}$. But $\mathrel{\mid\!\sim}_{\beta,\alpha} = \mathrel{\mid\!\sim}_\alpha$. It follows that $\mathrel{\mid\!\sim}_\alpha \beta$. This says that $\alpha \mathrel{\mid\!\sim} \beta$. We have proved that $\mathrel{\mid\!\sim}$ is monotonic. ∎

It follows from the above lemma that in a consequence revision system containing nonmonotonic consequence relations we have non-commuting revision operators, i.e. 'uncertainty relations'.

LEMMA 8.43. *Let $\mathcal{L} = \langle \mathcal{C}, F, \rightsquigarrow \rangle$ be a holistic logic such that every $\mathrel{\mid\!\sim}\, \in \mathcal{C}$ is monotonic. Then for every $\mathrel{\mid\!\sim}\, \in \mathcal{C}$ we have $\mathrel{\mid\!\sim} = \vdash_{\Sigma_{\mid\!\sim}}$, i.e. $\mathcal{L}$ is totally degenerate.*

Proof. Suppose that $\mathrel{\mid\!\sim}\, \in \mathcal{C}$ is monotonic. Let α be such that $\mathrel{\mid\!\not\sim} \alpha$. Hence $\mathrel{\mid\!\sim}_\alpha \neq \mathrel{\mid\!\sim}$. By 'provability of unprovability' we have that that $\mathrel{\mid\!\sim} \neg(\sigma \rightsquigarrow \alpha)$ and by monotonicity $\alpha \mathrel{\mid\!\sim} \neg(\sigma \rightsquigarrow \alpha)$. Since $[\neg(\sigma \rightsquigarrow \alpha)] = \{\mathrel{\mid\!\sim}, 0\}$, we have $\mathrel{\mid\!\sim}_\alpha = 0$. But this says that $\mathrel{\mid\!\sim} \neg\alpha$. It follows that for any α we have either $\mathrel{\mid\!\sim} \alpha$ or $\mathrel{\mid\!\sim} \neg\alpha$. We say that $\mathrel{\mid\!\sim}$ is complete as a consequence relation.

Recall that by $\Sigma_{\mid\!\sim}$ we denote the set $\{\alpha \mid \mathrel{\mid\!\sim} \alpha\}$. We have proved that for any α we have $\alpha \in \Sigma_{\mid\!\sim}$ or $\neg\alpha \in \Sigma_{\mid\!\sim}$. We prove, moreover, that $\Sigma_{\mid\!\sim}$ has the property that $(\alpha \rightarrow \beta) \in \Sigma_{\mid\!\sim}$ iff not $\alpha \in \Sigma_{\mid\!\sim}$ or $\beta \in \Sigma_{\mid\!\sim}$. It then follows that $\Sigma_{\mid\!\sim}$ is a complete theory, see the remark on lemma 1.15. Suppose that $(\alpha \rightarrow \beta) \in \Sigma_{\mid\!\sim}$. This says that $\neg(\alpha \wedge \neg\beta) \in \Sigma_{\mid\!\sim}$. By the property already proved this is equivalent to $\mathrel{\mid\!\not\sim} \alpha \wedge \neg\beta$. This is by a general condition imposed on the consequence relation considered the case iff $\mathrel{\mid\!\not\sim} \alpha$ or $\mathrel{\mid\!\not\sim} \neg\beta$ which in turn is equivalent to not $\alpha \in \Sigma_{\mid\!\sim}$ or $\beta \in \Sigma_{\mid\!\sim}$.

We now need to prove that $\mathrel{\mid\!\sim} = \vdash_{\Sigma_{\mid\!\sim}}$. For this we need to see that $\alpha \mathrel{\mid\!\sim} \beta$ iff $\mathrel{\mid\!\sim} \neg\alpha$ or $\mathrel{\mid\!\sim} \beta$. We have $\alpha \mathrel{\mid\!\sim} \beta$ iff $\mathrel{\mid\!\sim}_\alpha \beta$. Note that $\mathrel{\mid\!\sim}_\alpha = 0$, which means $\mathrel{\mid\!\sim} \neg\alpha$ or $\mathrel{\mid\!\sim}_\alpha = \mathrel{\mid\!\sim}$. $\mathrel{\mid\!\sim}_\alpha \beta$ therefore holds iff $\mathrel{\mid\!\sim} \neg\alpha$, which says that not $\alpha \in \Sigma_{\mid\!\sim}$ or $\mathrel{\mid\!\sim} \beta$, which means that $\beta \in \Sigma_{\mid\!\sim}$ ∎

The following theorem is the limiting case theorem for holistic logics.

THEOREM 8.44. *Let $\mathcal{L} = \langle \mathcal{C}, F, \rightsquigarrow \rangle$ be a holistic logic. Then the following conditions are equivalent.*

- *(i) Every $\mathrel{\mid\!\sim}\, \in \mathcal{C}$ is monotonic.*

- *(ii) For every $\mathrel{\mid\!\sim}\, \in \mathcal{C}$ we have $\mathrel{\mid\!\sim} = \vdash_{\Sigma_{\mid\!\sim}}$.*

- *(iii) All operators commute.*

Proof. (i) implies $(ii$ by lemma 8.43. That (ii) implies (iii) from the fact that if (ii) holds, then the operation is trivial in the sense that $F(\alpha, \mathrel|\!\sim) = \mathrel|\!\sim$ or $F(\alpha, \mathrel|\!\sim) = 0$. From 8.42 we see that (iii) implies (i). ∎

In the limit we have in particular monotonicity, classical consistency and thus *models*, a 'reality outside the logic'.

CHAPTER 9

TOWARDS HILBERT SPACE

Let us in this chapter come back to an idea which is central to the enterprise of this book. In chapter 6 we asked the question what logic could do about quantum mechanics. We said that it would not be our aim to find a new deductive system especially suited for reasoning in quantum mechanics, which does not mean that we would not consider this a reasonable enterprise. Rather we said that we should look for logical structures implicit in the formalism of quantum mechanics which could prove useful in the task of trying to understand this very formalism. We isolated two types of (related) structures: M-algebras and holistic logics. Both are abstractions from structures we find in Hilbert space. In this chapter we go further. We ask the question "Can we characterise the core concept of the formalism, namely that of a Hilbert space, in terms of these structures?" This is part of what in the literature on the foundations of quantum mechanics is sometimes called the representation enterprise. In this chapter we give, amongst other things, a characterisation of the concept of a classical Hilbert space in terms of logic.

1 Presenting holistic logics

1.1 The concept of a Hilbert space logic

Let H be an orthomodular space. Recall that $Sub(H)$ denotes the the set of closed subspaces of H. We know that $\langle Sub(H), \subset, ^\perp \rangle$ is an orthomodular lattice. Recall that $^\perp$ means orthogonal complement formation. We will, as we did earlier, use capital letters $A, B, ...$ for closed subspaces and, if there is no danger of confusion, for the corresponding projections. Moreover, we use the symbols for Boolean connectives in connection with closed subspaces, i.e we write $A \wedge B$ for $A \cap B$ and we denote the smallest closed subspace containing the closed subspaces A and B by $A \vee B$.

Let Fml be a propositional language as in chapter 1. Let $\Psi : Fml \to Sub(H)$ be a surjective function such that $\Psi(\neg \alpha) = \Psi(\alpha)^\perp$ and $\Psi(\alpha \wedge \beta)) = \Psi(\alpha) \wedge \Psi(\beta)$. Denote the projection corresponding to $\Psi(\alpha)$ by A. Let $x \in H$. Then we define the consequence relation $\vdash_x$ by

$$\alpha \vdash_{x, \Psi} \beta \text{ iff } Ax \in \Psi(\beta).$$

We will simply write $\vdash_x$ if Ψ is clear from the context. Note that $\vdash_x$ depends only on the ray of x, i.e. $\vdash_{x_1} = \vdash_{x_2}$ iff the one dimensional subspace $\langle x_1 \rangle$ generated by x_1 is equal to the one dimensional subspace $\langle x_2 \rangle$ generated by x_2.

Given an orthomodular space H and a function Ψ as described above, we define

$$\mathcal{C}_{H,\Psi} =: \{\vdash_x | \, x \in H\}.$$

Let us now define a function that will turn out to be an action on $\mathcal{C}_{H,\Psi}$. Define $\mathcal{F}_{H,\Psi} : Fml \times \mathcal{C}_{H,\Psi} \to \mathcal{C}_{H,\Psi}$ by

$$\mathcal{F}_{H,\Psi}(\alpha, \vdash_x) =: \vdash_{Ax}.$$

Note that $\mathcal{F}_{H,\Psi}$ is well defined, since $\langle x_1 \rangle = \langle x_2 \rangle$ implies $\langle Ax_1 \rangle = \langle Ax_2 \rangle$. Recall that the Sasaki hook $\leadsto_s$ is the connective defined as follows: $\alpha \leadsto \beta =: \neg \alpha \vee (\alpha \wedge \beta)$.

THEOREM 9.1. *Let H be an orthomodular space and Ψ a function as described above. Then $\mathcal{L}_{H,\Psi} =: \langle \mathcal{C}_{H,\Psi}, \mathcal{F}_{H,\Psi}, \leadsto_s \rangle$ is a holistic logic. All consequence relations satisfy the conditions we impose on consequence relations with the possible exception of Cut and Cautious Monotonicity. In case H is a Hilbert space all conditions are satisfied.*

The following proof is in case that H is a Hilbert space. Cut and Cautious Monotonicity work in the Hilbert space case only.

Proof. We first need to verify the conditions imposed on the elements of $\mathcal{C}$. This is routine for the most part.
Reflexivity is a consequence of the fact that for $x \in \Psi(\alpha)$ we have $Ax = x$.
Let us first verify *Cut*. So let $x \in H$ and assume $\alpha \wedge \beta \vdash_x \gamma$ and $\alpha \vdash_x \beta$. $\alpha \vdash_x \beta$ says that $\Psi(\alpha)_x \in \Psi(\beta)$. Moreover, from the above assumptions it follows that $\Psi(\alpha \wedge \beta)_x = \Psi(\alpha)_x$. By the hypothesis we have $\Psi(\alpha \wedge \beta)_x \in \Psi(\gamma)$ and thus $\Psi(\alpha)_x \in \Psi(\gamma)$. But this means that $\alpha \vdash_x \gamma$. Thus, *Cut* is verified. We now verify *Restricted Monotonicity*. Assume $\alpha \vdash_x \beta$ and $\alpha \vdash_x \gamma$. It follows that $\Psi(\alpha)_x = \Psi(\alpha \wedge \beta)_x$ and, since by the hypothesis we have $\Psi(\alpha)_x \in \Psi(\gamma)$, we see that $\Psi(\alpha \wedge \beta)_x \in \Psi(\gamma)$, which says that $\alpha \vdash_x \gamma$. Thus *Restricted Monotonicity* is verified.

In order to verify the other conditions use that by definition we have $\Psi(\alpha \wedge \beta) = \Psi(\alpha) \wedge \Psi(\beta)$ and $\Psi(\neg \alpha) = \Psi(\alpha)^\perp$ and elementary Hilbert space theory.

For the first global condition for instance suppose $\alpha \hspace{0.1em}\vert\hspace{-0.3em}\sim_{\mathcal{C}_{H,\Psi}} \gamma$ and $\beta \hspace{0.1em}\vert\hspace{-0.3em}\sim_{\mathcal{C}_{H,\Psi}} \gamma$. This means $\Psi(\alpha) \subset \Psi(\gamma)$ and $\Psi(\beta) \subset \Psi(\gamma)$. It is then elementary Hilbert space theory that $\Psi(\alpha \vee \beta) \subset \Psi(\gamma)$. But this says that $\alpha \vee \beta \hspace{0.1em}\vert\hspace{-0.3em}\sim_{\mathcal{C}_{H,\Psi}} \gamma$.

We now prove that $\mathcal{F}_{H,\Psi}$ is an action on $\mathcal{C}$. Condition (*i*) in the definition of an action is obvious, see definition 8.1. Consider condition (*ii*) in the

1. PRESENTING HOLISTIC LOGICS

definition of an action. Suppose $\vdash_x \neg \alpha$. This is equivalent to $x \in \Psi(\alpha)^\perp$, which is the case iff $Ax = 0$. But this means $\vdash_{Ax} = \mathcal{F}_{H,\Psi}(\alpha, \vdash_x) = 0$.
As to condition *(iii)* in the definition of an action let $\mathcal{F}_{H,\Psi}(\beta, (\mathcal{F}_{H,\Psi}(\alpha, \vdash_x)) = \mathcal{F}_{H,\Psi}(\alpha), \vdash_x)$. This is the case iff $BAx = Ax$, which is equivalent to $Ax \in \Psi(\beta)$. But this says that $\alpha \vdash_x \beta$.

We still need to prove that $\leadsto_s$ is internalising for $\mathcal{C}$. Suppose $\alpha \vdash_x \beta$. By definition this means that $Ax \in \Psi(\beta)$. By Proposition 2.26 this is the case iff $x \in \neg A \lor (A \land B)$. But this says $\vdash_x \alpha \leadsto_s \beta$.

We still need to see that $\mathcal{L}_{H,\Psi}$ is holistic. We need to show that any $\vdash_x$ has a pointer. Since Ψ is assumed to be surjective, it is easily seen that such a pointer is given by any formula σ such that $\Psi(\sigma) = \langle x \rangle$. ∎

We call a logic of the above form an *orthomodular space logic*. In case case H is a classical Hilbert space we call $\mathcal{L}_{H,\Psi}$ a *Hilbert space logic*.

1.2 The canonical $\mathcal{H}$-Model for a Hilbert space logic

We now give the construction of the canonical $\mathcal{H}$-model (see 8.17) for a given Hilbert space logic.

DEFINITION 9.2. Let H be a Hilbert space, Ψ a function as described in the last section and $x \in H$. Define the binary relation $\leq_x$ on H as follows

$$x_1 \leq_x x_2 \text{ iff: } d(x, x_1) \leq d(x, x_2)$$

Moreover, define the structure

$$\mathcal{M}_{x,\Psi} = \langle H, \leq_x, l_\Psi \rangle ,$$

as follows. Let $x \in H$, then $l_\Psi(x) = \{s_x\}$ is the singleton consisting of the following Scott-model s_x: For $\alpha \in Fml$ put $s_x(\alpha) = 1$, if $x \in \Psi(\alpha)$, else $s_x(\alpha) = 0$.

LEMMA 9.3. *Let $\mathcal{L}_{H,\Psi}$ be a Hilbert space logic. Then for every $x \in H$, $\mathcal{M}_{x,\Psi}$ is a GKLM model for $\vdash_{x,\Psi}$.*

Proof. We first have to verify the smoothness condition. For this observe that for any α we have $[\alpha] = \Psi(\alpha)$. Note that the notation $[\alpha]$ is in the sense of definition of a $(GKLM)$. It suffices to show that every $[\alpha]$ has a unique $\leq_x$-minimal element. But this is what the projection theorem 2.5 says, namely Ax is that unique minimal element.
It remains to be shown that $\vdash_{x,\Psi} = \vdash_{\mathcal{M}_{x,\Psi}}$. So let $\alpha \vdash_{x,\Psi} \beta$. By definition this means $Ax \in [\beta]$. But this is equivalent to $\alpha \vdash_{x,\Psi} \beta$, since Ax is the minimal element of $[\beta]$. ∎

We now define an $\mathcal{H}$-structure for a given Hilbert space logic $\mathcal{L}_{H,\Psi}$.

DEFINITION 9.4. Given the Hilbert space logic $\mathcal{L}_{H,\Psi} = \langle \mathcal{C}_{H,\Psi}, \mathcal{F}_{H,\Psi}, \leadsto_s \rangle$. Consider the structure $\mathcal{H}_{H,\Psi} = \langle H, h, \mathcal{F}, l_\Psi, g \rangle$ such that

- $h(x) = \vdash_x$
- $\mathcal{F}(\alpha, x) = Ax$
- The function l_Ψ is defined as follows: $l_\Psi(x) = \{s_x\}$, where $s_x(\alpha) = 1$ if $x \in \Psi(\alpha)$, 0 else.
- $g(x) = \leq_x$ as defined in Definition 9.2

THEOREM 9.5. *Given a Hilbert space logic* $\mathcal{L}_{H,\Psi} = \langle \mathcal{C}_{H,\Psi}, \mathcal{F}_{H,\Psi}, \leadsto_s \rangle$. *Then* $\mathcal{H}_{H,\Psi}$ *as defined above is an* $\mathcal{H}$-*model for* $\mathcal{L}_{H,\Psi}$.

1.3 Classical inconsistency in Hilbert space logics

We come back to the phenomenon first observed by Kochen and Specker, namely that Birkhoff-von Neumann quantum logic is 'classically inconsistent'. We will see that this phenomenon is not accidental. In fact, it is a consequence of the no windows theorem for holistic logics.

We start with the following simple observation. We denote by H_n an n-dimensional Hilbert space. Let x_1, x_2 be non-orthogonal and non-colinear vectors of H_2. Let Fml be the language of propositional logic and consider a Hilbert space logic $\mathcal{L}_{H_2,\Psi_0}$ such that for the propositional variables p_1, p_2 we have $\Psi_0(p_i) = \langle x_i \rangle$, $i = 1, 2$. Consider the formula $\phi = \phi_1 \wedge \phi_2 \wedge \phi_3 \wedge \phi_4$ such that

$$\phi_1 = p_1 \vee p_2$$

$$\phi_2 = \neg p_1 \vee p_2$$

$$\phi_3 = p_1 \vee \neg p_2$$

$$\phi_4 = \neg p_1 \vee \neg p_2$$

It is easily seen that ϕ is a classical contradiction which is provable in all consequence relations of $\mathcal{L}_H, \Psi_0$.

PROPOSITION 9.6. ϕ *is a classical contradiction and for all consequence relations* $\vdash$ *of* $\mathcal{L}_{H_2}, \Psi_0$ *we have* $\vdash \phi$

In the case of three dimensional Hilbert space H the above result is more difficult to establish. In [34] Kochen and Specker gave a classical tautology the negation of which is provable in all consequence relations of a Hilbert space logic presented by a three dimensional Hilbert space.

2. SYMMETRY AND HILBERT SPACE PRESENTABILITY

PROPOSITION 9.7 (Kochen-Specker).
 There exists a classical contradiction α and a Hilbert space logic $\mathcal{L}_{H_3,\Psi}$ such that $\not\hspace{-2pt}\sim \alpha$ for all $\sim$ of $\mathcal{L}$.

Remark: The formula presented by Kochen and Specker contains 117 variables. It represents the full space under a certain valuation' of these variables. It is important to note that this valuation is such that only compatible elements of $Sub(H_3)$ are combined by the connectives. The following theorem predicts the existence of a classical contradiction which is a 'quantum tautology'. It does, however, not capture the additional property of the Kochen-Specker formula just mentioned.

THEOREM 9.8. *Let H be a finite dimensional orthomodular space and $\dim H \geq 2$. Let $\mathcal{L}_{H,\psi}$ be a logic presented by H. Then Σ_g is classically inconsistent.*

COROLLARY 9.9. *Under the above hypotheses there exists a classical tautology ϕ such that for all $x \in H$ we have $\vdash_x \neg \phi$.*

Proof. The claim is an immediate consequence of theorem 9.1 and the global no windows theorem 8.41. Recall that the Sasaki hook is classically equivalent to material implication and note that holistic logics presented by finite-dimensional orthomodular spaces are finite dimensional as holistic logics. ∎

2 Symmetry and Hilbert space presentability: A representation theorem

In this section we are looking for properties characterising Hilbert space logics. To pose the problem more precisely, let us introduce the following terminology. Given a logic $\mathcal{L} = \langle \mathcal{C}, F, \leadsto \rangle$, a Hilbert space H and a function $\Psi \to Sub(H)$ such that $\mathcal{L} = \mathcal{L}_{H,\psi}$. Then we say that $\mathcal{L}$ is *presented* by H via Ψ. We say that $\mathcal{L}$ is presentable by H if there exists a function Ψ such that $\mathcal{L}$ is presented by H via Ψ. It is our aim to characterise the logics presentable by some Hilbert space H. In other words, we are looking for necessary and sufficient conditions for a logic $\mathcal{L}$ to be presentable by some Hilbert space H. We will see that, besides some natural logical conditions, there are two properties essential for the characterisation we have in mind. The first property is *holicity*. The second essential property is a symmetry property. We will call it the *symmetry property*. We will see that these two properties, namely *holicity* and *symmetry*, play a vital role in characterising Hilbert space logics.

LEMMA 9.10. *Let $\mathcal{L} = \langle \mathcal{C}, F, \leadsto_s \rangle$ be a holistic logic. Then $\langle Fml, \leq, ^* \rangle$ and thus $\langle Prop, \subset, ^* \rangle$ are orthomodular, atomistic and irreducible lattices.*

Proof. We have orthomodularity by the fact that $\mathcal{L}$ is a logic with the Sasaki hook as its internalising connective and theorem 8.13. As to atomicity observe that the atoms of $\langle Prop, \subset, ^* \rangle$ are of the form $[\sigma_{\hspace{-2pt}\sim}]$.

For irreducibility we need to prove that the centre of that lattice consists of truth and falsity only. For this it suffices to prove that for every proposition $[\alpha]$ not representing truth or falsity there exists an atom $[\sigma_{\hspace{-2pt}\sim}]$ such that $[\alpha]$ and $[\sigma_{\hspace{-2pt}\sim}]$ are not compatible. In the special case of a pointer $\sigma_{\hspace{-2pt}\sim}$ and a formula α compatibility says that $[\sigma_{\hspace{-2pt}\sim}] \subset [\alpha]$ or $[\sigma_{\hspace{-2pt}\sim}] \subset [\neg\alpha]$. Since $\mathcal{L}$ is non-trivial, for a given formula α there exists a $\hspace{-2pt}\sim_o \in \mathcal{C}$ such that neither $[\sigma_{\hspace{-2pt}\sim_0}] \subset [\alpha]$ nor $[\sigma_{\hspace{-2pt}\sim_0}] \subset [\neg\alpha]$ and thus $[\alpha]$ and $[\sigma_{\hspace{-2pt}\sim_0}]$ are not compatible. ∎

DEFINITION 9.11. Let $\mathcal{L} = \langle \mathcal{C}, F, \leadsto \rangle$ be a logic.

- We say that $\mathcal{L}$ has the upward finiteness property, in brief the uf-property, iff the following holds: Given a set Σ of formulas. Then there exists a formula ψ such that $\sigma \hspace{2pt}\mid\hspace{-4pt}\sim_\mathcal{C} \psi$ for every $\sigma \in \Sigma$ and the following condition is satisfied. For any formula ρ such that $\sigma \hspace{2pt}\mid\hspace{-4pt}\sim_\mathcal{C} \rho$ for every $\sigma \in \Sigma$, we have $\psi \hspace{2pt}\mid\hspace{-4pt}\sim_\mathcal{C} \rho$.

- We say that $\mathcal{L}$ has the downward finiteness property, in brief the df-property iff the following holds:

  Given a set Σ of formulas. Then there exists a formula χ such that $\chi \hspace{2pt}\mid\hspace{-4pt}\sim_\mathcal{C} \sigma$ for every $\sigma \in \Sigma$ and the following condition is satisfied. For any formula ρ such that $\rho \hspace{2pt}\mid\hspace{-4pt}\sim_\mathcal{C} \sigma$ for every $\sigma \in \Sigma$, we have $\rho \hspace{2pt}\mid\hspace{-4pt}\sim_\mathcal{C} \chi$.

- In case that $\mathcal{L}$ is holistic we say that $\mathcal{L}$ has the covering property iff the following condition is satisfied. Given a formula α and $\hspace{2pt}\mid\hspace{-4pt}\sim \in \mathcal{C}$ such that $\not\hspace{2pt}\mid\hspace{-4pt}\sim \alpha$. Then for any formula ρ such that $\alpha \hspace{2pt}\mid\hspace{-4pt}\sim_\mathcal{C} \rho$ and $\rho \hspace{2pt}\mid\hspace{-4pt}\sim_\mathcal{C} \alpha \vee \sigma_{\hspace{-2pt}\sim}$ we have $\rho \equiv_\mathcal{C} \alpha \vee \sigma_{\hspace{-2pt}\sim}$ or $\rho \equiv_\mathcal{C} \alpha$

Intuitively we may think of the formulas ψ and χ in the above definition of playing the role of 'infinite disjunction' and 'infinite conjunction' of the formulas of Σ. The properties defined above are such that the following lemma holds.

LEMMA 9.12. *Let $\mathcal{L} = \langle \mathcal{C}, F, \leadsto_s \rangle$ be a holistic logic having the df, uf and the covering properties. Then the lattices $\langle \overline{Fml}, \leq, ^* \rangle$ and thus $\langle Prop, \subset, ^* \rangle$ are orthomodular, atomic, irreducible, complete lattices having the covering property.*

Recall the following observation already made in 2.35.

Let H be a Hilbert space and let $(x_i)_{i \in I}$ be a complete orthonormal system of H. Then any permutation of the system $(x_i)_{i \in I}$, more precisely any permutation of the index set I, induces a unique unitary transformation

2. SYMMETRY AND HILBERT SPACE PRESENTABILITY 137

on H and thus an automorphism of the lattice $Sub(H)$. This fact reflects a symmetry property of Hilbert spaces and in view of Solér's theorem seems to be at the heart of the concept of a Hilbert space. It is the above fact that serves us as a motivation for the concept of a symmetric logic which we will study in the sequel.

DEFINITION 9.13. Let $\mathcal{L}$ be a holistic logic having the properties in the last lemma. Let $\Delta = (\mathrel{\mid\kern-0.4em\sim}_i)_{i \in I}$ be an infinite family of consequence relations of $\mathcal{L}$ with the following properties

- (i) For $i \neq j$, $\mathrel{\mid\kern-0.4em\sim}_i$ and $\mathrel{\mid\kern-0.4em\sim}_j$ are orthogonal.

- (ii) For any consequence relation $\mathrel{\mid\kern-0.4em\sim}$ of $\mathcal{L}$ there exists an $i_0 \in I$ such that $\mathrel{\mid\kern-0.4em\sim}$ and $\mathrel{\mid\kern-0.4em\sim}_{i_0}$ are not orthogonal.

Then we call Δ a basis for $\mathcal{L}$.

Remark: Intuitively, we may think of a basis Δ of a holistic logic $\mathcal{L}$ as follows. Given any consequence relation $\mathrel{\mid\kern-0.4em\sim}$ of $\mathcal{L}$. Then there exists a member of Δ in which $\mathrel{\mid\kern-0.4em\sim}$ is encoded. The system Δ may thus be viewed as containing the 'whole information' of $\mathcal{L}$.

DEFINITION 9.14. Let $\mathcal{L}$ be a logic as in the last definition and let $\Delta = (\mathrel{\mid\kern-0.4em\sim}_i)_{i \in I}$ be a basis for $\mathcal{L}$. We say that $\mathcal{L}$ satisfies the symmetry condition with respect to Δ iff the following holds. Let $f : I \to I$ be any permutation of the index set I. Then there exists an automorphism φ_f of the algebra of propositions of $\mathcal{L}$ (and thus of th algebra of operators) such that

- $\varphi_f([\sigma_i]) = [\sigma_{f(i)}]$, where $(\sigma_i)_{i \in I}$ is any family such that σ_i is a pointer to $\mathrel{\mid\kern-0.4em\sim}_i$.

- If the subset $J \subset I$ of those elements of I that are left fixed by f is non-empty, then φ_f induces the identity on $[0, A]$, where A is the smallest proposition containing $[\sigma_j]$ for all $j \in J$.

We say that $\mathcal{L}$ satisfies the (synonymously: is symmetric) iff there exists a basis Δ for $\mathcal{L}$ such that $\mathcal{L}$ is symmetric with respect to Δ.

Recall the notation $[0, A]$. It is the set of all propositions smaller than or equal to A equipped with a lattice structure in a natural way. In the following theorem we assume the 'presenting' function Ψ to be surjective. The next theorem is our *Representation Theorem*. The proof we give can be simplified by making use of the theorem 2.35 classical Hilbert lattices. Essentially, we repeat the argument in the proof of 2.35 so as to make this section as self-contained as possible.

THEOREM 9.15. *Let $\mathcal{L} = \langle \mathcal{C}, F, \leadsto_s \rangle$ be a logic. Then the following conditions are equivalent.*

- *(i) $\mathcal{L}$ is symmetric.*

- *(ii) There exists an infinite-dimensional classical Hilbert Space H presenting $\mathcal{L}$.*

Proof. For the direction from (ii) to (i) assume that there exists an infinite-dimensional classical Hilbert space H and a (surjective) function Ψ such that $\mathcal{L} = \mathcal{L}_{H,\Psi}$. We need to verify the symmetry property. Let $(x_i)_{i \in I}$ be a complete orthonormal system of H. Then $\Delta = (\vdash_{x_i})_{i \in I}$ is a basis for $\mathcal{L}$. Now observe that the lattice of propositions of $\mathcal{L}$ and $Sub(H)$ are isomorphic in a canonical way, namely via $[\alpha] \mapsto \Psi(\alpha)$. Thus, for the proof of symmetry it suffices to establish the following. For any permutation $f : I \to I$ there exists an automorphism ρ_f of $Sub(H)$ with the following properties:

- $\rho(\langle x_i \rangle) = \langle x_{f(i)} \rangle$

- If the set $J = \{i \mid f(i) = i\}$ is non-empty, then ρ_f induces the identical map on $[0, X]$, where X denotes the smallest closed subspace of H containing $\langle x_j \rangle$ for all $j \in J$.

To verify the above, recall that any for any $x \in H$ we have $x = \sum_{i \in I} \langle x, x_i \rangle x_i$. Define the map φ_f as follows. For $x = \sum_{i \in I} \langle x, x_i \rangle x_i$ put $\varphi_f(x) = \sum_{i \in I} \langle x, x_{f^{-1}(i)} \rangle x_i$. φ_f is well defined. We have for any $i \in I$ that $\varphi_f(x_i) = x_{f(i)}$. Moreover, φ_f is unitary, since for any $x, y \in H$ we have $\langle \varphi_f(x), \varphi_f(y) \rangle = \sum_{i \in I} \langle x, x_{f^{-1}(i)} \rangle \overline{\langle y, x_{f^{-1}(i)} \rangle} = \sum_{i \in I} \langle x, x_i \rangle \overline{\langle y, x_i \rangle} = \langle x, y \rangle$. Now assume that the set J of those elements which are left fixed by f is not empty. Denote by X the smallest closed subspace of H containing x_j for all $j \in J$. X is the smallest closed subspace containing $\{\langle x_j \rangle \mid j \in J\}$ and φ_f induces the identity on X. For the latter claim observe that φ_f induces the identity on the subspace spanned by $\{x_j \mid j \in J\}$ and X is the (topological) closure of that subspace. By continuity φ_f induces the identity on X too. Now, φ_f induces an ortholattice automorphism ρ_f on $Sub(H)$ such that for any $i \in I$, $\rho_f(\langle x_i \rangle) = \langle x_{f(i)} \rangle$. It is also evident that ρ_f induces the identical map on $[0, X]$. Thus the symmetry condition is verified.

For the other direction note that the existence of a basis guarantees that the lattice of propositions denoted by $Prop_\mathcal{L}$ has infinite height and observe that by Piron's representation theorem 2.28 there exists an orthomodular space H and an isomorphism $\Phi : Prop_\mathcal{L} \to Sub(H)$. We now exploit the symmetry property of $\mathcal{L}$ to prove that H must be a classical (infinite-dimensional) Hilbert space. Let Δ be a basis for $\mathcal{L}$ with respect to which

2. SYMMETRY AND HILBERT SPACE PRESENTABILITY

$\mathcal{L}$ is symmetric. Let $(\sigma_i)_{i \in I}$ be a corresponding family of pointers. We look at the family $\Phi([\sigma_i)])_{i \in I}$. Put $\langle x_i \rangle = \Phi([\sigma_i])$. This is an infinite pairwise orthogonal family of one dimensional subspaces (rays) of H. We will construct a family $(y_i)_{i \in I}$ of pairwise orthogonal elements of H such that for any $i, j \in I$ we have $\langle y_i, y_i \rangle = \langle y_j, y_j \rangle$. Then it follows by Solèr's theorem that H is a classical Hilbert space.

Let $i_0 \in I$ be fixed. Then for every $j \in I, j \neq i_0$ consider the permutation f_j of I defined as follows.

$$f_j(i_0) = j \text{ and } f_j(j) = i_0, \ f_j(i) = i \text{ else.}$$

The symmetry condition then guarantees that for every $j \in I, j \neq i_0$ there exists an automorphism φ_j of $Sub(H)$ such that

$$\varphi_j(\langle x_{i_0} \rangle) = \langle x_j \rangle$$

and, moreover, induces the identity on $[0, X]$, where X is the smallest closed subspace of H containing $\langle x_i \rangle$ for $i \neq i_0, j$. Clearly X has dimension greater than 2. In fact, it is infinite-dimensional. Mayet's theorem 2.33 then yields that φ_j is induced by some unitary operator ρ_j of H. For $j \neq i_0$ put $y_j = \rho_j(x_{i_0})$ and $y_{i_0} = x_{i_0}$. Since ρ_j is unitary and the $\langle x_i \rangle$'s are pairwise orthogonal, the family $(y_j)_{j \in I}$ is as required in Solèr's theorem 2.30 and H must be an infinite-dimensional classical Hilbert space.

We still need to prove that H presents $\mathcal{L}$. For this we first need to define the function Ψ. Define $\Psi : Fml \to Sub(H)$ by $\Psi(\alpha) = \Phi([\alpha])$. It is routinely verified that Ψ satisfies the conditions required.
We need to show

- 1. $\mathcal{C} = \mathcal{C}_{H,\Psi}$

- 2. If $\models = \vdash_x$, then for any α, $\models_\alpha = \mathcal{F}_{H,\Psi}(\alpha, \vdash_x)$

For 1. let $\models \in \mathcal{C}$ be given. We need to find a $\vdash_x \in \mathcal{C}_{H,\Psi}$ such that $\models = \vdash_x$. Let σ be a pointer to $\models$ and $\langle x \rangle = \Phi([\sigma]) = \Psi(\sigma)$. We have $\alpha \models \beta$ iff $\sigma \models \alpha \leadsto_s \beta$ iff $[\sigma] \subset [\alpha \leadsto_s \beta]$. This is equivalent to $\langle x \rangle \subset \Psi(\alpha \leadsto_s \beta)$ which says $\alpha \vdash_x \beta$. Thus $\models = \vdash_x$. For a given $\vdash_x$ the same reasoning applies to finding a $\models \in \mathcal{C}$ such that $\vdash_x = \models$.
For 2. let $\models = \vdash_x$. Note that $\beta \models_\alpha \gamma$ iff $\alpha \models (\beta \leadsto_s \gamma)$ iff $\alpha \vdash_x (\beta \leadsto_s \gamma)$ iff $\beta \vdash_{Ax} \gamma$. But $\vdash_{Ax} = \mathcal{F}_{H,\Psi}(\alpha, \vdash_x)$. ■

Remark: The reader may have noticed that in he above proof we did not use the second condition of the definition of a basis. In fact the argument works without that condition. If we omit the second condition we can no longer say that Δ contains 'the whole information' of $\mathcal{L}$. Instead, its intuitive function would be to guarantee that $\mathcal{L}$ is 'rich in information' in that it contains infinitely many non-orthogonal consequence relations, which thus are not encoded in each other.

3 Formal reflections on the connectives in Hilbert space logics

This section is strongly technical in nature. The reader can skip it without missing anything essential from the conceptual point if view.

Quantum logic, however it may be defined, is certainly one of those branches of logic in which the connectives are least understood. In this section we take a closer look at the problem of the connectives. This is another application of the local viewpoint. For a given Hilbert space H and some $x \in H$ we look at the consequence relation $\vdash_x$ as defined and studied in the previous sections. The main observation underlying the considerations of this section is that, as pointed out by Lehmann in [39], it makes sense to study those consequence relations in the absence of connectives, i.e. for a (poor) language without connectives. Moreover, we may in place of the consequence relation $\vdash_x$, which is a binary relation between formulas, study the more general concept of an inference operation $\mathcal{C}_x$, which is a binary relation between a set of formulas $\mathcal{A}$ and and a formula β with $\beta \in \mathcal{C}_x(\mathcal{A})$ having the intuitive meaning "β is a consequence of the set of formulas $\mathcal{A}$".

So we start with a language without connectives and then study conservative extensions of the inference operation to a language containing connectives.

Given a Hilbert space H and $x \in H$. We will see that this situation gives rise to an inference operation $\mathcal{C}_x$, which we call a quantum inference operation, on a language without connectives. We then study conservative extensions of $\mathcal{C}_x$ to languages containing connectives and discuss their properties from various points of view.

3.1 Quantum inference operations

Given a non-empty set $\mathcal{P}$, which we regard as a set of atomic propositions. We assume $\mathcal{P}$ to contain $\top$ and $\bot$, the symbol for logical truth and falsity respectively. Thus the language we start with has no connectives. Given a Hilbert space H, an element x of H and a function $\Psi : \mathcal{P} \to Sub(H)$ such that $\Psi(\top) = H$ and $\Psi(\bot) = \{0\}$. We assume Ψ to be surjective. As usual we occasionally write A for $\Psi(\alpha)$ as well as for the projector corresponding to $\Psi(\alpha)$. We then have the consequence relation $\vdash_x$ over $\mathcal{P}$ presented by Ψ:

$$\alpha \vdash_x \beta \text{ iff } Ax \in B$$

3. FORMAL REFLECTIONS ...

As observed in [39], this definition can be extended to the definition of an inference operation as defined in chapter 1 as follows. Let $\mathcal{A}$ be any set of (atomic) formulas. We may then define what it means to say that β is a consequence of $\mathcal{A}$. Namely, consider $A =: \bigcap \{\Psi(\alpha) \mid \alpha \in \mathcal{A}\}$. Note that A is again a closed subspace of H so that we may define β to be a consequence of $\mathcal{A}$, i.e. $\beta \in C(\mathcal{A})$, iff $A x \in B$.

Denote the set of x-consequences of $\mathcal{A}$ by $\mathcal{C}_x(\mathcal{A})$. Context permitting we omit the subscript x. These inference operations called quantum inference operations have the following properties as is routinely checked.

$$\text{For any } \mathcal{A} \subset \mathcal{P} \text{ we have } \mathcal{A} \subset C(\mathcal{A}) \text{ Inclusion}$$

and

$$\mathcal{A} \subset \mathcal{B} \subset C(\mathcal{A}) \text{ implies } C(\mathcal{A}) = C(\mathcal{B}) \text{ Cumulativity}$$

In [39] inference operations satisfying the above conditions are called C-logics. C-logics in general admit a smooth characterisation in terms of a representation theorem (see [39]) and are worth studying in their own right. Keep in mind that quantum consequence operations are C-logics.

Consider now the closure L of $\mathcal{P}$ under the unary connective $\neg$ and the binary connective $\wedge$. The other connectives are defined as usual in terms of $\neg$ and $\wedge$. We will for a given inference operation $\mathcal{C}_x$ study conservative extensions of $\mathcal{C}_x$.

Consider the following three conditions concerning the connectives $\wedge$ and $\neg$.

$\wedge$-R- $\mathcal{C}(A, \alpha \wedge \beta) = \mathcal{C}(A, \alpha, \beta)$

$\neg$ R1- $\mathcal{C}(A, \alpha, \neg\alpha) = L$

$\neg$ R2- if $\mathcal{C}(A, \neg\alpha) = L$, then $\alpha \in \mathcal{C}(A)$.

Call a set $\mathcal{A}$ of formulas a theory if $\mathcal{C}(\mathcal{A}) = \mathcal{A}$. Call $\mathcal{A}$ consistent if $\mathcal{C}(\mathcal{A})$ is not the full language. Given an inference operation $\mathcal{C}$ over $\mathcal{P}$ and let $\mathcal{C}'$ be an extension of $\mathcal{C}$ to the language L or some other language extending $\mathcal{P}$. Then we say that $\mathcal{C}'$ is a *conservative* extension of $\mathcal{C}$ if for any $\mathcal{A} \subset \mathcal{P}$ we have $\mathcal{C}'(\mathcal{A}) \cap \mathcal{P} = \mathcal{C}(\mathcal{A})$.

In the following considerations we need the concept of a proof operator.

DEFINITION 9.16. Given a consequence relation $\mid\sim$. Suppose the language contains a connective (primitive or definable by other connectives) such that we have

- $\hspace{0.5em}\vdash \alpha$ iff $\hspace{0.2em}\vdash \Box \alpha$
- $\not\vdash \alpha$ iff $\hspace{0.2em}\vdash \neg\Box\alpha$

Then we call $\Box$ a proof operator for $\vdash$.

3.2 The Birkhoff-von Neumann Extension

We first consider the following extension of Ψ to L.

$$\Psi(\alpha \wedge \beta) = \Psi(\alpha) \cap \Psi(\beta)$$

$$\Psi(\neg \alpha) = \Psi(\alpha)^{\perp}$$

Note that Ψ goes into $Sub(H)$, even onto, and the extended consequence relation and inference operation is defined as follows. Given a set $\mathcal{A}$ of formulas of L and $\alpha \in L$. Put $A = \bigcap \{\Psi(\beta) \mid \beta \in \mathcal{A}\}$. We say that $\alpha \in \mathcal{C}_x^{BvN}$ iff $Ax \in \Psi(\alpha)$. If there is no danger of confusion, we omit the subscript x. This extension is in the spirit of [4] in that we invoke orthogonal complement formation in Hilbert space in order to present negation. We refer to it as the Birkhoff-von Neumann extension.

3.3 The Engesser-Gabbay Extension

The following extension was, essentially, first considered in [17].

In the next definition we use the following abbreviation: $\Box \alpha =: \sigma_x \leadsto_s \alpha$ and $\sigma_x \leadsto_s \alpha =: \neg \sigma_x \vee (\sigma_x \wedge \alpha)$.

We consider the following language.

DEFINITION 9.17. Given a set $\mathcal{P}$ of atomic formulas. Then define the language $\mathcal{L}^{EG}$ by the following clauses.

- (i) If $\alpha \in \mathcal{P}$, then $\Box \alpha \in L^{EG}$.

- (ii) If $\varphi \in L^{EG}$, then $\Box \varphi \in L^{EG}$.

- (ii) If $\varphi, \psi \in L^{EG}$, then $\neg \varphi, \varphi \wedge \psi \in L^{EG}$.

  The other connectives are defined as usual. Define the language $\mathcal{L}^{EG}$ as $\mathcal{P} \cup L^{EG}$.

Note that $\mathcal{L}^{EG}$ is a sublanguage of L. A typical formula of $\mathcal{L}^{EG}$ is for instance $\Box \alpha \to \neg \Box \neg \alpha$ with $\alpha \in L$. Extend Ψ as in the Birkhoff-von Neumann extension and and let $\sigma_x \in \mathcal{P}$ be such $\Psi(\sigma_x) = \langle x \rangle$.

DEFINITION 9.18. We define $\mathcal{C}_x^{EG}$ (Engesser-Gabbay extension) as follows. Given $x \in H$, $\mathcal{A} \subset \mathcal{L}^{EG}, \varphi \in \mathcal{L}^{EG}$. Let $A = \bigcap \{\Psi(\alpha \mid \alpha \in \mathcal{A}\}$. Then define $\varphi \in \mathcal{C}_x^{EG}(\mathcal{A})$ iff $Ax \in \Psi(\varphi)$.

3. FORMAL REFLECTIONS ...

3.4 The Lehmann Extension

In [39] the following result is proved.

THEOREM 9.19 (Lehmann). *Let $\mathcal{C}$ be a $\mathcal{C}$-logic over $\mathcal{P}$. Then there exists a $\mathcal{C}$-logic $\mathcal{C}'$ over $\mathcal{L}$ which is a conservative extension of $\mathcal{C}$ and which satisfies $\wedge - R, \neg - R1, \neg - R2$.*

Since the inference operations $\mathcal{C}_x$ are $\mathcal{C}$-logics, it follows from the above theorem that they admit a conservative extension as described in the theorem.

In [39] an explicit construction for such a conservative extension is given. We now describe this construction in the special case of a quantum consequence operation. In this formal presentation it may be the case that the intuitions behind the construction are not as transparent as would be desirable. For a deeper understanding of this construction we refer the reader to [39]. Given $y \in H$, put $T_y =: \{A \in Sub(H) \mid y \in A\}$.

The following result characterises the consistent theories of a quantum consequence operation.

THEOREM 9.20. *Let $\mathcal{A} \subset \mathcal{P}$. Then $\mathcal{A}$ is an x-consistent theory iff there is a y not orthogonal to x such that $\Psi(\mathcal{A}) = T_y$.*

Proof. For the direction from left to right put $A =: \bigcap\{\Psi(\mathcal{A}) \mid A \in \mathcal{A}\}$ and $Ax =: y$. Then $\langle y \rangle \in \Psi(\mathcal{C}(\mathcal{A}))$ since Ψ is assumed to be surjective. Since $\mathcal{C}(\mathcal{A}) = \mathcal{A}$, we have have $\Psi(\mathcal{C}(\mathcal{A})) = \Psi(\mathcal{A})$ and $\langle y \rangle \in \Psi(\mathcal{A})$. Thus $A \subset \langle y \rangle$. Since A is assumed to be consistent and $\langle y \rangle$ is one-dimensional, we have $A = \langle y \rangle$. It follows that $\Psi(\mathcal{A}) \subset T_y$.

Now assume assume $B \in T_y$. Note there exists a $\beta \in \mathcal{P}$ such that $B = \Psi(\beta)$. We have $Ax \in B$. This says that $\beta \in \mathcal{C}(\mathcal{A})$. Again, since $\mathcal{C}(\mathcal{A}) = \mathcal{A}$, we have $B \in \Psi(\mathcal{A})$. It follows that $T_y \subset \Psi(\mathcal{A})$. We have now established that $\Psi(\mathcal{A}) = T_y$.

It remains to be shown that y is not orthogonal to x. Suppose to the contrary that $y \in \langle x \rangle^\perp$, i.e. $Ax \in \langle x \rangle^\perp$. This means that $Ax = 0$ and thus $\mathcal{A}$ is inconsistent contrary to the hypothesis.

For the other direction it suffices to show that $\mathcal{C}(T_y) \subset T_y$. Let $B \in \mathcal{C}(T_y)$. It follows that the projection operator corresponding to $\langle y \rangle$ applied to x is in B. We then have $\langle y \rangle \subset B$ and thus $B \in T_y$. ∎

The above theorem says that there is a one-to-one correspondence between the consistent theories of the quantum inference operation $\mathcal{C}_x$ and the elements of H not orthogonal to x.

Given a quantum inference operation $\mathcal{C}_x$ over $\mathcal{P}$ presented by the function $\Psi : \mathcal{P} \to Sub(H)$. Given $S \subset H$, then by S^c we mean the complement of S in H. For a given $x \in H$ consider $H_x := (\langle x \rangle^\perp)^c$ and define the function

$\Psi_L : \mathcal{P} \to 2^{H_x}$ presenting the Lehmann extension as follows. For $\alpha \in \mathcal{P}$ define

$$\Psi_L(\alpha) = \Psi(\alpha) \cap H_x,$$

In view of 9.20, the set H_x represents the set of consistent theories of the inference operation.

For the connectives define

$$\Psi_L(\neg \alpha) = (\Psi_L(\alpha))^c,$$

$$\Psi_L(\alpha \wedge \beta) = \Psi_L(\alpha) \cap \Psi_L(\beta).$$

Let A be any subset of H_x. Then we denote by A^* the smallest closed subspace of H containing A and as always, if there is no danger of confusion, the corresponding projection operator. For $A \in Sub(H)$ we put $A^x = (A \cap H_x)^*$. We now define $\mathcal{C}^L$ omitting the subscript x.

DEFINITION 9.21. Let $\mathcal{C}_x$ be a quantum inference operation over $\mathcal{P}$ presented by the function Ψ. We define the Lehmann extension $\mathcal{C}^L$ as follows. Let $\mathcal{A}$ be any set of formulas of $\mathcal{L}$. Then define $\mathcal{C}^L(\mathcal{A})$ as follows. Let β be any formula of $\mathcal{L}$. Then consider $S =: \bigcap \{\Psi^L(\alpha) \mid \alpha \in \mathcal{A}\}$. Now, if $S^*x \in S$, we say that $\beta \in \mathcal{C}^L(\mathcal{A})$ if $S^*x \in \Psi^L(\beta)$. If not $S^*x \in S$ we define $\beta \in \mathcal{C}^L(\mathcal{A})$ if $S \subset \Psi^L(\beta)$.

This definition is the result of applying the general construction as given in the proof of Theorem 2 in [39] to quantum inference operations.

LEMMA 9.22. Let $A \in Sub(H)$ and $x \in H$. Then $Ax \in \langle x \rangle^\perp$ implies $Ax = 0$.

Proof. Assume $Ax \in \langle x \rangle^\perp$. This means that $\langle Ax, x \rangle = 0$. Note that $AA = A$. It follows that $\langle AAx, x \rangle = 0$. Since A is self-adjoint, we get $\langle Ax, Ax \rangle = 0$ and thus $Ax = 0$. ∎

LEMMA 9.23. Let $A \subset Sub(H)$ and suppose $Ax \neq 0$. Then $Ax = A^x x$.

Proof. First observe that $A^x \subset A$. Moreover, note that, if $Ax \neq 0$, then we have by the last lemma that not $Ax \in \langle x \rangle^\perp$ and thus $Ax \in A \cap H_x$, i.e. $Ax \in A^x$. Since Ax and $A^x x$ are the unique elements of A and A^x respectively which are closest to x, it follows that $Ax = A^x x$. ∎

The following obvious lemma is useful.

LEMMA 9.24. Let $S \subset Sub(H)$ and let $A = \bigcap \{B \cap H_x \mid B \in S\}$. Then $A^* = (\bigcap \{B \mid B \in S\})^x$.

THEOREM 9.25. $\mathcal{C}^L$ is a C-Logic and a conservative extension of $\mathcal{C}_x$.

3. FORMAL REFLECTIONS ...

Proof. Let us verify that $\mathcal{C}^L$ is in fact a C-logic.

So given sets of formulas $\mathcal{A}, \mathcal{B}$ such that $\mathcal{A} \subset \mathcal{B} \subset \mathcal{C}^L(\mathcal{A})$. We then need to prove that $\mathcal{C}^L(\mathcal{A}) = \mathcal{C}^L(\mathcal{B})$. Let $S =: \bigcap \{\Psi^L(\alpha \mid \alpha \in \mathcal{A}\}$ and $T =: \bigcap \{\Psi^L(\beta) \mid \beta \in \mathcal{B}\}$. We have to consider four cases.
Case 1: $S^*x \in S$ and $T^*x \in T$.
Case 2: $S^*x \in S$, but not $T^*x \in T$.
Case 3: not $S^*x \in S$, but $T^*x \in T$.
Case 4: not $S^*x \in S$ and not $T^*x \in T$.
We prove case 1. Note that $T \subset S$. We have for every $\beta \in \mathcal{B}$ that $S^*x \in \Psi^L(\beta)$. It follows that $S^*x \in T$ and thus $S^*x = T^*x$. Now let $\beta \in \mathcal{C}^L(\mathcal{B})$. This means $T^*x \in \Psi^L(\beta)$ and thus $S^*x \in \Psi^L(\beta)$ which means that $\beta \in \mathcal{C}^L(\mathcal{A})$.
We prove case 2. First note that $S^*x \in \Psi(\beta)$ for all $\beta \in \mathcal{B}$ and thus $S^*x \in T$. Let $\beta \in \mathcal{C}^L(\mathcal{B})$. Then we have $T \subset \Psi^L(\beta)$. It follows that $S^*x \in \Psi^L(\beta)$.
We prove case 3. In this case $\mathcal{B} \subset \mathcal{C}^L(\mathcal{A})$ implies $S \subset T$. Since $T \subset S$, we have $S = T$. If $\beta \in \mathcal{C}^L(\mathcal{B})$, we have $T^*x = S^*x \in \Psi(\beta)$, hence $\beta \in \mathcal{C}^L(\mathcal{A})$.
We prove case 4. In this case we, again, have $S = T$. $\beta \in \mathcal{C}^L(\mathcal{B})$ says $T \subset \Psi^L(\beta)$. Thus $S \subset \Psi^L(\beta)$, which means $\beta \in \mathcal{C}^L(\mathcal{A}$.

We now prove that $\mathcal{C}^L$ is an extension of $\mathcal{C}_x$. Suppose $\beta \in \mathcal{C}_x(\mathcal{A})$. Put $S =: \bigcap \{\Psi^L(\alpha) \mid \alpha \in \mathcal{A}\}$ and $A =: \bigcap \{\Psi(\alpha) \mid \alpha \in \mathcal{A}\}$. We have $Ax \in \Psi(\beta)$. Assume $Ax \neq 0$. It follows by lemma 9.23 and lemma 9.24 that $A^x x = Ax = S^*x \in S$. Thus $S^*x \in \Psi^L(\beta)$. This is the first case in Definition 9.21 and we have $\beta \in \mathcal{C}^L(\mathcal{A})$. Now suppose $Ax = 0$. It follows that S is empty. So $S \subset \Psi^L(\beta)$. We have the second case in Definition 9.21 and $\beta \in \mathcal{C}^L(\mathcal{A})$.

We now prove that $\mathcal{C}^L$ is a conservative extension of $\mathcal{C}_x$. For this we need to show that for any set $\mathcal{A} \subset \mathcal{P}$ and any $\beta \in \mathcal{P}$ $\beta \in \mathcal{C}^L(\mathcal{A})$ implies $\beta \in \mathcal{C}_x(\mathcal{A})$. Assume $\beta \in \mathcal{C}^L(\mathcal{A})$. Assume the first case in definition 9.21, namely that $S^*x \in S$. Hence $\beta \in \mathcal{C}^L(\mathcal{A})$ iff $S^*x \in \Psi^L(\beta)$. Since $S^*x \neq 0$, we have $S^*(x) = A^x(x) = Ax$ by lemma 9.23. It follows that $A(x) \in \Psi^L(\beta) \subset \Psi(\beta)$, which says that $\beta \in \mathcal{C}_x \mathcal{A}$. Now assume the second case of definition 9.21, i.e. not $S^*x \in S$. This means that $S^*x \in \langle x \rangle^\perp$. This is, by lemma 9.23 possible only if $S^x x = 0$ and thus, by lemma 9.22 $Ax = 0$. It follows that $\beta \in \mathcal{C}_x(\mathcal{A})$. ∎

3.5 Discussing the extensions

Let us now compare the extensions introduced in the last section. The Birkhoff-von Neumann extension is the richest. Its salient features are the following. First, negation, which is defined via orthogonal complement formation, does not satisfy $\neg - R2$. Second, it admits an internalising connective definable in terms of $\neg$ and $\wedge$, namely the Sasaki hook. Third, it is (classically) inconsistent in the sense that for any x there exists a classical contradiction φ such that $\varphi \in \mathcal{C}^{BvN}(\top)$. In fact, such a contradiction can be chosen uniformly for any x, say the negation of the Kochen-Specker tautology if the dimension of H is at least 3. Moreover, the Birkhoff-von Neumann extension admits a proof operator definable in terms of $\neg$ and $\wedge$, which is the source of self-referential soundness and completeness.

We will prove that the Lehmann extension is classically consistent. It satisfies $\neg - R2$. However, it does not admit an internalising connective definable in terms of $\neg$ and $\wedge$ and it does not admit a definable proof operator.

THEOREM 9.26. *The Birkhoff-von Neumann extension is a conservative extension of $\mathcal{C}_x$ which is classically inconsistent. It admits an internalising connective and a proof operator both definable by $\neg$ and $\wedge$. Negation does not satisfy $\neg - R2$.*

Proof.
We have conservativity by definition. The fact that the Birkhoff-von Neumann extension is classically inconsistent is the Kochen-Specker phenomenon ('no windows) with which we dealt several times, see chapters 6, 8 and this chapter. The Sasaki hook $\leadsto_s$ is an internalising connective of the sort required. The proof operator is given by $\Box \alpha := \sigma_x \leadsto_s \alpha$, where σ_x is any formula such that $\Psi(\sigma_x = \langle x \rangle)$. The only claim we don't know yet is that the Birkhoff- on Neumann extension does not satisfy $\neg R2$. A counterexample is provided by the following geometrical fact. Consider three mutually non-parallel and non-orthogonal lines A, B, C through the origin in the real plane. Let x be a vector of C. Let α, β, γ be formulas such that $\Psi(\alpha) = A$, $\Psi(\beta) = B$, $\Psi(\gamma) = C$. Then we have in slight abuse of notation that $\mathcal{C}_x(\alpha, \neg \beta) = L$. But not $\beta \in \mathcal{C}_x(\alpha)$ since the projection of x onto A is not in B. ∎

In the following theorem we write $\varphi \vdash_x^{EG} \psi$ for $\psi \in \mathcal{C}(\{\varphi\})$.

THEOREM 9.27. *The extension $\mathcal{C}^{EG}$ is a conservative extension of $\mathcal{C}$ which is classically consistent. In fact $\mathcal{C}^{EG}(\top)$ is consistent. The set $\{\varphi \in L^{EG} \mid \varphi \in \mathcal{C}^{EG}\}$ even forms a complete classical theory. For any $\varphi, \psi \in \mathcal{L}^{EG}$ we have $\varphi \vdash_x^{EG} \psi$ iff $\vdash_x^{EG} \sigma_x \leadsto_s (\varphi \leadsto_s \psi)$. For $\varphi, \psi \in L^{EG}$ we have $\varphi \vdash_x^{EG} \psi$ iff $\vdash_x^{EG} \varphi \to \psi$. $\Box$ is a proof operator which is trivial in its action on formulas of L^{EG}. The EG-extension satisfies $\neg - R2$ for formulas of L^{EG}.*

Proof. Again we have conservativity by definition. Let us verify classical consistency. We need to prove that $\Sigma =: \mathcal{C}(\top)$ is classically consistent. For this we construct a model for Σ, i.e. a valuation V such that $V(\varphi) = 1$ for all $\varphi \in \Sigma$. Define V as follows. For $\alpha \in L$ we put $V(\alpha) = 1$ if $x \in \Psi(\alpha)$, else $V(\alpha) = 0$ and extend it truth-functionally as usual in classical propositional

3. FORMAL REFLECTIONS ...

logic. It is then readily verified that $V(\Box\varphi) = 1$ iff $x \in \Psi(\varphi)$. Considering that Hilbert space logics are holistic and in view of proposition 8.29 we moreover see that $V(\varphi) = 1$ iff $x \in \Psi(\varphi)$. But $x \in \Psi(\varphi)$ means $\varphi \in \mathcal{C}^{EG}$. We have proved that the Engesser-Gabbay extension is classically consistent.

Now note that the formulas of L^{EG} satisfy the hypothesis of proposition 8.29. That $\varphi \vdash_x^{EG} \psi$ is equivalent to $\vdash_x^{EG} \varphi \to \psi$ is then (v) of that proposition. That $\neg - R2$ is satisfied is a consequence of 8.29 too. ∎

Let us now study the Lehmann extension more closely.
The following lemma is an immediate consequence of definition 9.21.

LEMMA 9.28.

- (i) $\alpha \wedge \beta) \in \mathcal{C}^L(\{\top\})$ iff $\alpha \in \mathcal{C}^L(\{\top\})\}$ and $\beta \in \mathcal{C}^L(\{\top\})$
- (ii) $\neg \alpha \in \mathcal{C}^L(\{\top\})$ iff not $\alpha \in \mathcal{C}^L(\{\top\})$

LEMMA 9.29. *Suppose* $\mathcal{C}^L(\mathcal{A}) = L$. *Then* $S =: \bigcap \{\Psi^L(A) \mid A \in \mathcal{A}\} = \emptyset$

Proof. By the hypothesis we have in particular that $\bot \in \mathcal{C}^L(\mathcal{A})$. We have $\Psi^L(\bot) = \emptyset$. We therefore cannot have the first case in definition 9.21. So we must have the second case in definition 9.21. But this says that $S = \emptyset$. ∎

THEOREM 9.30. $\mathcal{C}^L$ *is a conservative extension of* $\mathcal{C}_x$ *which is classically consistent. Negation satisfies* $\neg - R2$. *It admits no internalising connective definable by* $\neg$ *and* $\wedge$ *and no proof operator definable by* $\neg$ *and* $\wedge$.

Proof. We give a sketch of the proof. We leave it to the reader to make the argument fully formal, which is tedious but, in principle, not difficult.

First we prove that $\mathcal{C}^L$ satisfies $\neg - R2$. Suppose that $\mathcal{C}^L(\mathcal{A}, \neg \alpha) = L$. By lemma 9.29 we have that the intersection of $S =: \bigcap\{\Psi^L(A) \mid A \in \mathcal{A}\}$ and $\Psi^L(\alpha)$ is empty. It follows that $S \subset \Psi^L(\alpha)$ and by inspecting definition 9.21 we see that $\alpha \in \mathcal{C}^L(\mathcal{A})$.

In order to see that $\mathcal{C}^L$ is classically consistent note that in view of lemma 9.28 we can, for a given $x \in H$, assign a 'truth value' $V(\alpha)$ to every $\alpha \in \mathcal{L}$ in a natural way as follows. For atomic $\alpha \in \mathcal{P}$ define $V(\alpha) = 1$ iff $\vdash_x \alpha$ and V and extend V canonically to $\mathcal{L}$. Then we can prove, using induction, that $V(\alpha) = 1$ iff $\vdash_x \alpha$. This means that any α such that $\vdash_x \alpha$ is classically consistent.

Let us now see why the Lehmann extension admits neither an internalising connective definable by $\wedge$ and $\neg$ nor such a proof operator. Assume to the contrary that there exists an internalising connective $C(\alpha, \beta$ such that $\alpha \in \mathcal{C}(\beta))$ iff $\mathcal{C}(\top, C(\alpha, \beta)$. From this we see that whether $\beta \in \mathcal{C}(\alpha)$ holds would depend only on the truth values of α and β as defined above. Now, it easy to see geometrically, say in the real plane, that the following constellation is possible: not $\alpha \in \mathcal{C}^L(\top)$, $\beta \in \mathcal{C}^L(\top)$, $\gamma \in \mathcal{C}(\top)$, $\beta \in \mathcal{C}^L(\alpha$ and not $\gamma \in \mathcal{C}^L(\alpha)$. However, by 'truth-functionality' we should have $\beta \in \mathcal{C}^L(\alpha)$ and $\gamma \in \mathcal{C}^L(\alpha)$, which is a contradiction. A similar argument establishes the non-existence of a definable proof operator for the Lehmann extension. Namely suppose there exists a proof operator defined by a formula $C(\alpha, \beta, \gamma_1, ..., \gamma_n)$ with fixed $\gamma_1, ..., \gamma_n$. In this case the truth values of $\gamma_1, ..., \gamma_n$ are fixed and whether $\beta \in \mathcal{C}^L(\alpha)$ would depend only on the truth values of α and β, which is not the case as shown above. ∎

CHAPTER 10

FINAL REFLECTIONS

1 The plot of quantum mechanics: consistency, encodedness, nonmonotonicity

In this section we reflect on some of the crucial features of quantum mechanics. Although these reflections have their origin in the formal results of the previous chapters, we, in this final chapter, allow ourselves a less formal mode of reflection and even a touch of narrative literature in our style of writing.

As already pointed out several times, among the most salient properties of quantum mechanics unfamiliar from classical mechanics there are the following. First, in quantum mechanics, there are *uncertainty relations*, a phenomenon we do not have in classical mechanics. Second, in quantum mechanics there is, generally, a *change of state in measurement*. In classical mechanics there is no such change. Another interesting and widely discussed feature of quantum mechanics concerns the *holistic aspect of quantum reality*. As mentioned earlier, it is widely held that quantum mechanics suggests a revision of the fragmenting view of (physical) reality we are used to from classical mechanics in favour of a more holistic view. Those are the properties we want to focus on.

The above properties are important characters in the play of quantum mechanics. They do not act in isolation but there is a plot in which they are all, each in its own specific role, involved and interconnected. It is for instance plausible to assume a connection between the first two of the characters, namely the uncertainty relations on the one hand and change of state in measurement on the other. In fact, it is hard to imagine how there can be 'non-simultaneous measurability' without change of state in measurement. It seems that change of state in measurement is a necessary condition for the existence of uncertainty relations. The question arises whether this is the whole story or whether there are deeper connections still to be discovered. An interesting question for instance is whether we can relate the intuition of holism arising in quantum mechanics to the other properties mentioned. At the level of the quantum mechanical formalism such a connection is not obvious.

At a deeper level, however, such connections can in fact be revealed, as we will see. We will tell the plot of the play of quantum mechanics at the level of logic so to speak. We will see that, at that level, new characters, which do not appear on the platform of the formalism, will enter the stage and play an important role in the play. We then get a broader picture of the characters and their mutual relations.

Let us for the sake of the following discussion recall the essentials of chapter 8 so that those readers who haven't studied the technical details of the previous chapters can enjoy the story.

First, what, essentially, is a logic as defined in chapter 8?

What we, for the purposes of the last two chapters, called a logic is a triple of the form $\langle \mathcal{C}, F, \leadsto \rangle$. It has its formal motivation in a structure arising in a natural way in classical logic (see section 9.1).

The first ingredient of such a logic is a set $\mathcal{C}$ of logical entities called consequence relations. A consequence relation may, traditionally, be viewed as a logic in its own right. So, according to traditional terminology, a logic in this sense is a set of logics. Of course, all except one of these individual logics are required to be consistent, i.e. for any $\mathrel{\vert\!\sim} \in \mathcal{C}$ there is no α such that $\mathrel{\vert\!\sim} \alpha$ and $\mathrel{\vert\!\sim} \neg\alpha$. It has turned out that it is useful to allow for the inconsistent (universal) consequence relation. Another requirement is that these various logics that constitute a logic in our sense should not form too heterogeneous a collection. We may interpret the requirement that all the various consequence relations have a common internalising connective $\leadsto$ as expressing the intuition that $\mathcal{C}$ is not too heterogeneous. So the second ingredient is the internalising connective common to all consequence relations. We will in the sequel call the consequence relations the states of the logic.

The third ingredient is that of an action F on $\mathcal{C}$. Formulas *act* on $\mathcal{C}$, and in general, this action is proper, i.e. given a formula α and some $\mathrel{\vert\!\sim} \in \mathcal{C}$ we have in general that $\mathrel{\vert\!\sim}_\alpha =: F(\alpha, \mathrel{\vert\!\sim}) \neq \mathrel{\vert\!\sim}$. We think of the function F as a 'revision' function as follows. Given any $\mathrel{\vert\!\sim} \in \mathcal{C}$ and some formula α, then we view $F(\alpha, \mathrel{\vert\!\sim})$ also denoted by $\mathrel{\vert\!\sim}_\alpha$ as $\mathrel{\vert\!\sim}$ revised by α. Note that we have $\mathrel{\vert\!\sim}_\alpha$. Think intuitively of $\mathrel{\vert\!\sim}_\alpha$ as that consequence relation of $\mathcal{C}$ which proves α and is 'most similar' to $\mathrel{\vert\!\sim}$.

What are holistic logics?

A holistic logic is a logic having certain interesting properties which we will describe below.

The states of a holistic logic are — except one — *consistent*. They are moreover *nonmonotonic*. As indicated earlier, we regard nonmonotonicity as reflecting the presence of uncertainty relations in quantum mechanics at the level of logic. Revising a consequence relation generally also means

1. THE PLOT OF QUANTUM MECHANICS 151

changing it. We call this feature of our logics *revision*. We regard it as reflecting the quantum-mechanical phenomenon of (nonmonotonic) change of state. By nonmonotonic change of a state we mean that we cannot gain new knowledge in the transition from one state to another without at the same time losing knowledge.

Another crucial property of holistic logics is that their states *encode each other*. More precisely, any state of a holistic logic encodes any other state non-orthogonal to it. Note that 'encode' does not mean 'contain'. In fact, in a holistic logic no state contains any other state despite the fact that non-orthogonal states encode each other. As a special case of encodedness we have what we called self-referential soundness and completeness. Namely each consequence relation encodes itself. We regard this property of encodedness and particular that of self-encodedness as reflecting the intuition of holicity to which quantum mechanics gives rise.

Another salient property of certain holistic logics, namely those arising from Hilbert spaces, for instance, is expressed in the theorems we called no windows theorems. Essentially these theorems say that given any state $\mid\sim\, \in \mathcal{C}$ of such a (holistic) logic, then the set of statements proved by $\mid\sim$ has no (classical) model. It does not correspond to any classical reality.

Those, now, are the logical properties we want to reflect on: consistency, (self-)encodedness, nonmonotonicity, change of state (revision), and classical inconsistency.

These properties represent so to speak the characters interacting in the story we are going to tell. Note that at the level of logic a new character inevitably enters the stage, namely consistency. This property is not salient at the level of the quantum mechanical formalism. It will turn out, however, that at the logical level it is crucial to the understanding of the interplay of all the properties in question.

1.1 Feynman's logical tightrope: the uncertainty principle

The Feynman lectures on physics are remarkable in many respects. What is remarkable with regard to the present discussion is that Feynman seems to be the only author of a textbook on physics to use a typically logical terminology in explaining Heisenberg's uncertainty principle to the (beginning) student. In his discussion of the uncertainty principle Feynman constantly emphasises its connection with logic, namely with the logical property of consistency. According to Feynman the uncertainty principle serves a logical function.

Here are some quotations from Volume 3 of the Feynman lectures. Feynman writes: "Let us show for one particular case that the kind of relation given by Heisenberg must be true in order to keep quantum mechanics from getting into trouble." What does he mean by 'getting into trouble'? To get a clearer picture, here is another quotation: "The uncertainty principle protects quantum mechanics." What does the uncertainty principle 'protect' quantum mechanics from? Again a quotation: "...if a way to beat the uncertainty relation were ever discovered, quantum mechanics would give inconsistent results and would have to be discarded as a valid theory of nature". Now the keyword has been given: inconsistency! According to Feynman the uncertainty principle protects quantum mechanics from *inconsistency*. This is according to Feynman the connection between the uncertainty principle and and the logical notion of consistency. Another of Feynman's comments on the uncertainty principle: "This is the logical tightrope on which we must walk if we wish to describe nature successfully". Quantum mechanics, according to Feynman, is a way of describing nature which can be consistent only if the uncertainty principle holds. The uncertainty principle or the uncertainty relations in general are built into the system of quantum mechanics and are thus part of the system. But at the same time they constitute that part of the system that protects it from inconsistency.

Clearly, Feynman's remarks are to be taken as purely intuitive serving the pedagogical purpose of conveying a feeling for the uncertainty principle to the student. There is therefore no point in trying to find out what he precisely had in mind with his remarks. What, however, we can do is to attempt a reconstruction of Feynman's remarks in a logical framework regardless of what he actually had in mind.

We analyse the statement that the uncertainty principle protects quantum mechanics from inconsistency as a statement of the following form. Given a system S, say a logical system, of a certain type T and given a property P. What does it mean to say that the property P protects the system S from inconsistency? There are two possibilities of rephrasing this logically. The first possibility is to view the property P as a sufficient condition for the consistency of S. In this case we could rephrase as follows. Given a system S of type T with property P, then S is consistent. The other possibility is to view the property P as a necessary condition of the consistency of S. In this case the rephrasing would be: Given a system S of type T, then S can be consistent only if it has the property P. We will play the game with the latter version.

This is still very general and we have to make a commitment concerning the term system. We take a system to be a logic as described. Moreover, we know that it is reasonable to regard the property of nonmonotonicity as the logical counterpart of the uncertainty relations. Feynman's statement has then the following translation: Given a logic $\mathcal{L}$ having property T. Then $\mathcal{L}$

1. THE PLOT OF QUANTUM MECHANICS 153

is consistent only if it is nonmonotonic. The question is what is condition T. The point we will make is that here the property reflecting holicity comes into play in a natural way. We need, at this point, not care about what holicity means precisely. At any rate we will arrive at the following insight: If a logic is holistic, then it can be consistent only if it is nonmonotonic. In Feynman's terminology: Nonmonotonicity protects holistic logics from inconsistency.

1.2 How an agent with full introspection can be consistent

Let us now be serious about the question of how holicity plays together with consistency and nonmonotonicity. In particular we will have to say what holicity means. We said that in the case of a holistic logic in the precise sense of chapter 8 holicity is reflected as mutual encodedness of states. In particular we saw that in a holistic logic any state (consequence relation) encodes itself. If we view the states of a holistic logic as representing the states of an agent, then we may say that this agent has full introspection or is self-aware in all states. Formally this is expressed as the fact that every state admits a proof operator, see 9.16. Namely, given a state $\mathrel{\mid\!\sim}$ of a holistic logic with the pointer σ_x and the internalising connective $\rightsquigarrow$, then the proof operator (9.16) is given by $\Box\alpha = \sigma_x \rightsquigarrow \alpha$.

The simple and fairly obvious statement below — formulated as a proposition — expresses an elementary fact which, however, sheds light on the phenomena we intend to discuss and their interplay.

PROPOSITION 10.1. *Given a consequence relation* $\mathrel{\mid\!\sim}$. *Suppose* $\mathrel{\mid\!\sim}$ *admits an internalising connective* $\rightsquigarrow$ *and a proof operator* $\Box$. *Then* $\mathrel{\mid\!\sim}$ *has no proper consistent extension with* $\rightsquigarrow$ *as an internalising connective* $\Box$ *as a proof operator.*

COROLLARY 10.2. *Let* $\mathrel{\mid\!\sim}$ *be consistent having the proof operator* $\Box$ *and the internalising connective* $\rightsquigarrow$. *Suppose* $\mathrel{\mid\!\not\sim} \alpha$. *Assume* $\mathrel{\mid\!\sim}_0$ *is consistent such that* $\mathrel{\mid\!\sim}_0 \alpha$. *Then* $\mathrel{\mid\!\sim}_0$ *admits* $\Box$ *as a proof operator and* $\rightsquigarrow$ *as an internalising connective only if there exists a formula* β *such that* $\mathrel{\mid\!\sim} \beta$ *and not* $\mathrel{\mid\!\sim}_0 \beta$.

Note that the above theorem also holds in the limiting case of a consequence relation of the form $\vdash_\Sigma$ where Σ is a complete classical theory. The reason is that in classical logic the proof operator is trivial ('invisible') and a complete classical theory is also maximal consistent (see chapter 1). Note that the corollary expresses nonmonotonicity of change of state. In the case of a complete classical theory it doesn't say anything because in this case there is no state $\mathrel{\mid\!\sim}_0$ satisfying the hypothesis of the corollary. So nonmonotonicity is not implied in the classical case. The reader may view the argument in the proof of proposition 10.1 given below as a formalisation of the informal argument we gave for the claim that an autoepistemic reasoner with full introspection cannot be monotonic.

Proof. Let $\mathrel{|\!\sim}$ be as in the hypothesis. Suppose there exists a consistent $\mathrel{|\!\sim}_0$ which has $\leadsto$ as an internalising connective and $\square$ as a proof operator and which extends $\mathrel{|\!\sim}$ properly, i.e. we have $\mathrel{|\!\sim} \subset \mathrel{|\!\sim}_0$ and this inclusion is proper. This means that there exist formulas α and β such that $\alpha \mathrel{|\!\not\sim} \beta$ and $\alpha \mathrel{|\!\sim}_0 \beta$. It follows that $\mathrel{|\!\not\sim} \alpha \leadsto \beta$ and $\mathrel{|\!\sim}_0 \alpha \leadsto \beta$. We then have that $\mathrel{|\!\sim} \neg\square(\alpha \leadsto \beta)$ and since $\mathrel{|\!\sim}_0$ is an extension of $\mathrel{|\!\sim}$ that $\mathrel{|\!\sim}_0 \neg\square(\alpha \leadsto \beta)$. But on the other hand we have, since $\square$ is assumed to be a proof operator for $\mathrel{|\!\sim}_0$, that $\mathrel{|\!\sim}_0 \square(\alpha \leadsto \beta)$, which contradicts the consistency of $\mathrel{|\!\sim}_0$.

We get the corollary as follows. Assume that $\mathrel{|\!\sim}_0$ is consistent with the proof operator $\square$ and the internalising connective $\leadsto$. Further assume that the conclusion is false, i.e. there is no β such that $\mathrel{|\!\sim} \beta$ and not $\mathrel{|\!\sim}_0 \beta$. This would mean that $\mathrel{|\!\sim}_0$ is a consistent extension of $\mathrel{|\!\sim}$ with the proof operator $\square$ and the internalising connective $\leadsto$. Moreover, it would be a proper consistent extension with these properties, since $\mathrel{|\!\sim}_0 \alpha$ and not $\mathrel{|\!\sim} \alpha$. This, however, is impossible by the above theorem. Therefore there must exist a β such that $\mathrel{|\!\sim} \beta$ and not $\mathrel{|\!\sim}_0 \beta$. ∎

What now do the above observations mean in our picture of the self-aware agent. We require such an agent to be self-aware, i.e. having , in all his states. Formally this is expressed by the requirement that all states have a proof operator and an internalising connective. Second, our picture must allow for change. The agent must have the possibility of changing his state (of mind), i.e pass from one state of mind to another. If we think of the agent's states as states of knowledge, this transition may be viewed as a learning process. What the corollary says has then the following interpretation. More precisely, given any state $\mathrel{|\!\sim}$ and a formula α, then our picture must allow for the revision of $\mathrel{|\!\sim}$ by α, i.e the transition to a state $\mathrel{|\!\sim}_\alpha$ such that $\mathrel{|\!\sim}_\alpha \alpha$. What the corollary says is that in learning something new or becoming aware of something new the agent cannot be monotonic. The process of learning or becoming aware of something cannot be cumulative in the sense that the agent is now aware of more than he was aware of before. Rather the change of state must be such that he must have 'forgotten' something. He must have lost awareness of something he had been aware of before. Otherwise he couldn't be consistent. The nonmonotonic nature of the transition from one state to another keeps the self-aware agent consistent.

The point of the consideration is this. Self-awareness is a special case of the more general phenomenon of mutual encodedness of states which we

1. THE PLOT OF QUANTUM MECHANICS

said reflects the intuition of holicity. The above consideration then shows that such a 'holistic' agent can 'change his state of mind' consistently only in a nonmonotonic way. Nonmonotonicity is the price for maintaining both consistency and self-awareness.

Let us now come to another character in the plot, namely classical inconsistency. Recall that given a consequence relation $\mathrel{|\!\sim}$ we say that $\mathrel{|\!\sim}$ is classically inconsistent if the set of formulas $\Sigma_{|\!\sim}$ defined by $\Sigma_{|\!\sim} =: \{\alpha \mid \mathrel{|\!\sim} \alpha\}$ is classically inconsistent. Note moreover that the classical inconsistency of $\mathrel{|\!\sim}$ does not imply that $\mathrel{|\!\sim}$ is inconsistent as a consequence relation. In a Hilbert space logic for instance all non-zero consequence relations are consistent as consequence relations but classically inconsistent.

Now, how does classical inconsistency interact with the other characters?

PROPOSITION 10.3. *Suppose $\mathrel{|\!\sim}$ is nonmonotonic admitting $\rightarrow$ as an internalising connective and a proof operator $\Box$ which is definable in terms of the other connectives. Suppose further that there exists a set of formulas Σ such that $\mathrel{|\!\sim} \alpha$ implies $\Sigma \vdash \alpha$. Suppose $\Sigma \vdash \alpha$ implies $\Sigma \vdash \Box \alpha$. Then Σ is inconsistent.*

Proof. Let α and β be such that $\mathrel{|\!\sim} \beta$ and $\alpha \mathrel{|\!\not\sim} \beta$. Such formulas exist because $\mathrel{|\!\sim}$ is assumed to be nonmonotonic. Then we have

$$(1) \quad \Sigma \vdash \beta$$

By classical logic we get

$$(2) \quad \Sigma \vdash \alpha \rightarrow \beta$$

Hence

$$(3) \quad \Sigma \vdash \Box(\alpha \rightarrow \beta)$$

On the other hand we have, since $\alpha \mathrel{|\!\not\sim} \beta$

$$(4) \quad \mathrel{|\!\not\sim} \alpha \rightarrow \beta$$

and thus

$$(5) \quad \mathrel{|\!\sim} \neg\Box(\alpha \rightarrow \beta)$$

and

$$(6) \quad \Sigma \vdash \neg\Box(\alpha \rightarrow \beta)$$

So, (3) and (6) say that Σ is inconsistent. ∎

The above observation is the principle underlying the proofs of the no windows theorems.

First, essentially it says that a consequence relation which is nonmonotonic and self-encoded cannot be classically consistent. At first glance this might not seem too surprising. In fact, one might expect that the clash between the monotonicity of classical logic and the nonmonotonicity of $\mid\!\sim$ will suffice to produce the effect of classical inconsistency of $\mid\!\sim$. In principle this is true. There is an important subtlety here however. Namely, it is only in combination with the property of encodedness that the clash between the monotonicity of classical logic and the nonmonotonicity of $\mid\!\sim$ produces the effect of the classical inconsistency of $\mid\!\sim$. To see this clearly let us look at the argument of the above proof more closely.

In the argument it is (2) and (3) on the one hand and (5) and (6) on the other that produce the clash. Clearly, it is the monotonic nature of classical logic that gives us (2) and (3). In order to establish (5) and thus (6), however, nonmonotonicity of $\mid\!\sim$ is not sufficient. We also need self-encodedness.

To summarise, the effect to be explained, namely the inconsistency of Σ, is the result of the clash between the monotonicity of classical logic on the one hand and the fact that $\mid\!\sim$ is nonmonotonic *and* self-encoded. Once we have nonmonotonicity as a character and also self-encodedness as a character, then the character of classical inconsistency automatically enters the stage.

The above proposition is of course not as strong as the theorems we called no windows theorems. But its proof sheds light on the mechanism that brings about the effect of classical inconsistency also in the no windows theorems. The reader is advised to study the proofs of the no windows theorems with the above in mind.

The perceptive reader has probably realised that in the above considerations we tacitly assumed a uniform proof operator for all states. More precisely, we assumed the following. Given a state $\mid\!\sim$ and its revision by α, i.e. $\mid\!\sim_\alpha$. Then we assumed that both states have the same proof operator. This, however, is an oversimplification in view of the fact that this assumption is not true in the case of holistic logics in the sense of chapters 8 and 9. Take for instance a Hilbert space logic and consider $\vdash_x$ and $\vdash_y$ for distinct x and y. Then the corresponding proof operators $\Box_x$ and $\Box_y$ are given by $\Box_x \alpha = \sigma_x \leadsto_s \alpha$ and $\Box_y \alpha = \sigma_y \leadsto_s \alpha$, which are distinct. So the above considerations do not apply to the case of a holistic logic in the technical sense of chapters 9 and 10.

Actually, in the above considerations we were concerned with encodedness in the special sense of self-encodedness (self-awareness). The only thing we assumed was that every state encodes itself. Similar considerations are

1. THE PLOT OF QUANTUM MECHANICS

possible in the case of genuine mutual encodedness of states. We do not discuss this in detail. Let us just scratch the surface. This is enough to see that in such considerations we encounter another character in the plot, namely that of *symmetry*. So let us us assume now that every state has a proof operator which now, however, may depend on the state as is the case for holistic logics.

Given two distinct states $\mathrel{\vdash}_x$ and $\mathrel{\vdash}_y$ which are mutually encoded. Call the respective proof operators $\square_x$ and $\square_y$. In holistic logics encoding has the following properties.

Given a state x that encodes state y. Then we have

$$\forall \alpha (\mathrel{\vdash}_x \alpha \text{ iff } \mathrel{\vdash}_y \square_x \alpha)$$

$$\forall \alpha (\mathrel{\not\vdash}_x \alpha \text{ iff } \mathrel{\vdash}_y \square_x \neg \square_x \alpha)$$

It is now important to note that encoding is symmetric. If x encodes y, then y encodes x and we have

$$\forall \alpha (\mathrel{\vdash}_y \alpha \text{ iff } \mathrel{\vdash}_x \square_y \alpha)$$

$$\forall \alpha (\mathrel{\not\vdash}_y \alpha \text{ iff } \mathrel{\vdash}_x \square_y \neg \square_y \alpha)$$

This is how things are in the case of holistic logics for non-orthogonal x and y. Here note two things. First, the above says that $\mathrel{\vdash}_x$ encodes $\mathrel{\vdash}_y$ and $\mathrel{\vdash}_y$ encodes $\mathrel{\vdash}_x$. Second note that the encoding is symmetric in the sense that we get the way $\mathrel{\vdash}_x$ encodes $\mathrel{\vdash}_y$ by interchanging x and y and vice versa. Obviously we are dealing here not just with mutual encodedness of states but with *symmetric* mutual encodedness. As we will argue, non-monotonicity here comes not only from the interplay between consistency and encodedness but from the interplay of consistency, encodedness *and* symmetry (of encodedness).

It is therefore natural to require encodedness to be symmetric in the sense that given $\mathrel{\vdash}_x$ and $\mathrel{\vdash}_y$, then for any any feature of $\mathrel{\vdash}_x$ that is encoded in $\mathrel{\vdash}_y$ the 'symmetric' corresponding feature of $\mathrel{\vdash}_y$ is encoded in $\mathrel{\vdash}_x$.

Now we want the analogue of corollary 10.2 in case we allow the proof operators to depend on the state. It suffices to see that given two states which symmetrically encode each other, then neither of them can be a proper subset of the other. Assume that symmetry means that for any statement about encodedness the symmetric statement holds too. Suppose that $\mathrel{\vdash}_x$ is a proper subset of $\mathrel{\vdash}_y$. We will show this would violate the symmetry of encodedness as described. In fact it it is easily checked that

$$\forall \alpha (\mathrel{\not\vdash}_y \alpha \text{ implies } \mathrel{\vdash}_x \neg \square_x \alpha)$$

By symmetry we should now have

$$\forall \alpha (\not\hspace{-2pt}\sim_x \alpha \text{ implies } \sim_y \neg \Box_y \alpha)$$

In order to see that this does not hold take an α such that $\not\hspace{-2pt}\sim_x \alpha$ and $\sim_y \alpha$.

So the assumption that $\sim_x$ is a proper subset of $\sim_y$ would violate a natural principle of symmetric encodedness. In terms of our story this means that in case we allow for state-dependent proof operators it seems that encodedness alone is not sufficient for nonmonotonicity. If we introduce symmetry (of encodedness), however, as a new character, then nonmonotonicity enters the stage automatically.

Let us now give a summary of the plot. The characters of the plot are consistency, encodedness (holicity), nonmonotonicity, symmetry, change of state (revision), classical inconsistency. With some of these characters we are well acquainted from (the level of) quantum mechanics, with others we are not. In fact a few of them do not appear at all at the level of quantum mechanics but only at the logical level. We are well familiar with nonmonotonicity, which at the level of quantum mechanics appears in the guise of the uncertainty relations. We are also familiar with change of state which is expressed at the level of quantum mechanics as the projection postulate. In a sense we are also familiar with the character of encodedness. At the level of quantum mechanics, however, this character does not present itself as clearly as the other two. It is present only as an intuition revealing its nature only at a deeper level. Consistency is a character which is not salient at the level of quantum mechanics. It is logical in nature appearing only at the logical level. But at that level its appearance is natural. The other two characters, classical inconsistency and symmetry of encodedness, again, are not salient in quantum mechanics. Classical inconsistency is the Kochen-Specker phenomenon, a discovery by logicians. As to the character of symmetry of encodedness we must note that it does not represent what at the level of the quantum mechanical formalism is referred to as symmetry. So again, this character comes in only at the logical level. All these characters are intimately connected in the plot of quantum mechanics as described.

What are the essential relationships between the characters of the play?

We saw that once we have consistency and encodedness we must also have nonmonotonicity and in particular nonmonotonic change of state. Recall what Feynman says: The uncertainty relations (nonmonotonicity) are necessary for the consistency of quantum mechanics. Nonmonotonicity flows from consistency and encodedness. If the first two characters are present then the third must be present too for the plot to make sense. In a certain variation of the plot we saw that we need another character, namely

1. THE PLOT OF QUANTUM MECHANICS

symmetry of encodedness, for this connection. Why did the author of the play create the character of classical inconsistency? The answer is that this was a consequence of his having introduced consistency and encodedness. The fact that he introduced consistency and encodedness compelled him to introduce nonmonotonicity which in turn forced him to introduce classical inconsistency. What now is the leading character of the play? We sympathise with the answer that it is the character of encodedness, which represents the intuition of holicity. Its presence is vital for the explanation of the presence of the others except consistency. But consistency is a character present in any play on logic and can hardly be considered the main character in such a play. We see that holicity, which at the level of quantum mechanics plays the role of a vague intuition, only becomes an important character in the play at the logical level.

What is the relationship between the plot of quantum mechanics and the plot of classical mechanics? Again the character of encodedness provides the explanation for this relationship. In the plot of classical mechanics, encodedness is not present at all. Only consistency is present. So there was no need for the author to invent the others. The play of classical mechanics has only one character, namely consistency. It is a monologue.

1.3 The invisible proof operator in classical logic and classical mechanics

Let us now revisit the structures we considered in 9.1. Given a consistent set of formulas Σ.

PROPOSITION 10.4. *Let Σ be consistent. Then $\vdash_\Sigma$ admits a (definable) proof operator iff Σ is complete (and thus maximal consistent). In this case the proof operator is trivial, namely we have that $\Box \alpha$ is logically equivalent to α.*

Proof. Suppose $\vdash_\Sigma$ admits a definable proof operator $\Box$. Then we have by the definition of a proof operator and by classical logic that for any α, $\vdash_\Sigma \Box \alpha \leftrightarrow \alpha$ and thus $\vdash_\Sigma \neg \Box \alpha \to \neg \alpha$. Suppose that not $\Sigma \vdash \alpha$ which says that not $\vdash_\Sigma \alpha$. It follows that $\vdash_\Sigma \neg \Box \alpha$ and by the above remark $\vdash_\Sigma \neg \alpha$. This says that $\Sigma \vdash \neg \alpha$. Σ is thus complete. $\Box \alpha =: \top \to \alpha$ defines a (trivial) proof operator. ■

Intuitively, we may view the simple fact stated above in the following light. As we saw several times earlier in the book, families of the form $(\vdash_{\Sigma_i})_{i \in I}$, where Σ_i is a complete classical theory, arise as limiting cases of genuine (non-degenerate) holistic logics having a non-trivial proof operator. In the limiting case the proof operator applied to a formula α disappears.

$\Box \alpha$ 'collapses' to α. The statement "α is measurable" collapses to the assertion of α. Provability (measurability) becomes just truth. In the limiting case the proof operator becomes invisible so to speak. This explains the fact that in classical mechanics we need not care about measurability. Instead of saying $\Box \alpha$ meaning "α is measurable" we just assert α saying "α".

2 A speculative look at the measurement problem

2.1 General remarks

In this section we take, in the semi-formal and sketchy way adopted in this chapter, a look at the so-called measurement problem. In this we do not want to commit ourselves to any philosophical position nor do we claim to have a 'solution' to this notorious problem. Rather we investigate what resources we have in our logical framework for treating it.

The idea is simple and can be described as follows. Physicists use to describe their experimental setup in a measurement in classical terms. Bohr even famously insisted that this has to be so. It seems that in the process of measurement the measuring instrument, which generally is a macroscopic object, plays the role of a classical object obeying the laws of classical physics. Now, in our framework, 'classical' and 'quantum' reside side by side with 'classical' being a special and limiting case. In such a framework it is natural to treat measurement as an interaction of a classical system with a quantum system. We will play the game of measurement in our logical framework so to speak and see what we get irrespective of how things actually are in nature.

Before describing the measurement problem in the dramatic form of Schrödinger's cat let us explain its general nature and why it is crucial to the understanding of quantum mechanics and in particular its mathematical formalism. In quantum mechanics the concept of measurement plays a vital role - in contrast to classical mechanics. Recall that it is one of the basic principles of quantum mechanics that, given an observable A represented mathematically as the Hermitian operator (again called) A, then the eigenvalues of A are those quantities that we can find as values of the observable A when a *measurement* of A is performed. The uncertainty relations are statement concerning 'non-simultaneous measurability' of certain observables. Generally, quantum mechanics makes statements about the outcome of measurements.

Now, the process of measurement is a physical process itself and we may expect quantum mechanics to give us an adequate description of this process. This, however, is apparently not the case. Essentially, the measurement problem or the measurement paradox, as it is perhaps more adequately called, consists in the fact that on the hand the formalism of quantum me-

2. A SPECULATIVE LOOK ...

chanics is about the outcome of measurements but on the other hand seems to give incorrect results when applied to the process of measurement itself.

What are the characteristics of measurement in physics? Measurement is a physical process involving the interaction of two systems, the system to be measured and the measuring system also called the measuring instrument. The systems interact in such a way that one of the interacting systems, namely the measuring instrument, indicates the value of a certain observable pertaining to the system to be measured. This applies equally to quantum mechanics and classical physics. Clearly, not every interaction between two systems constitutes a measurement. There are of course physical interactions between systems in which neither 'measures' the other. Rather, in practice, the measuring instrument is a macroscopic object with a scale or a screen from which the result of measurement can be read off by the human eye. Bohr famously insisted that the entire experimental arrangement even in the case of quantum measurement must be described in terms of classical physics.

In principle, the formalism of quantum mechanics permits us to treat any interaction of two systems. Both are, in the formalism, represented by their corresponding Hilbert spaces the tensor product of which represents the composite system. But how is the particular nature of the measurement process reflected in the formalism? In particular, how are the different roles of the system to be measured and the measuring instrument respectively reflected in the formalism? What, if anything, is special about the Hilbert spaces of measuring instruments? As far as we see, these questions have no answer within the formalism of quantum mechanics.

The issue of measurement in quantum mechanics is closely linked to another issue, namely that of the 'collapse of the wave function' or (synonymously) the projection postulate. Recall that the projection postulate says the following. Suppose a measurement of an observable A represented by an Hermitian operator (denoted again by) A is performed. Then after measurement the system is in an eigenstate of A and the corresponding eigenvalue is the value of A measured. It is important to note this link between measurement, which is a physical process, and the phenomenon of 'collapse'. It is not just that we experience the strange phenomenon of collapse (projection) in quantum mechanics but we have to bear in mind that it is in the process of measurement that it occurs.

We may therefore expect a theory of measurement to *explain* this phenomenon rather than to presuppose it.

2.2 The measurement problem in a nutshell

We view the measurement problem in the dramatic version of Schrödinger's cat. Given a quantum system, say an electron. We want to measure the z-component (or any other component) of the spin of the electron. It is known that this observable can assume only two values: "up" and "down". We assume that the experimental arrangement of the measurement process is as follows. In the case "spin = down" some device is triggered which kills a cat. If we have "spin = up" the cat stays alive. The cat is thus the measuring instrument.

Assume that before measurement the electron is (as a fermion) in the singlet state

$$(1)\quad 1/\sqrt{(2)}(|\,up\rangle - |\,down\rangle)$$

Call the system consisting both of the cat, the measuring instrument, and the electron, the system to be measured, the composite system. Then the formalism of quantum mechanics tells us that, after measurement, the composite system is in the following state:

$$(2)\quad 1/\sqrt{(2)}(|\,alive\rangle \otimes |\,up\rangle - |\,dead\rangle \otimes |\,down\rangle)$$

So, after measurement the state of the composite system is an entanglement. This is at odds with the facts, if we insist that after measurement the cat is either alive and spin up or the cat is dead and spin down, both with probability 1/2. The formalism, however, says that the z-component of spin is not sharp nor is the cat's life. This is what in the literature is often referred to as: "The cat is half alive and half dead". Such a state has never been observed. Rather the system is either in state

$$|\,alive\rangle \otimes |\,up\rangle$$

or in state

$$|\,dead\rangle \otimes |\,down\rangle$$

Both states have probability 1/2.

Therefore, it seems that the formalism of quantum mechanics does not provide the correct prediction when applied to the measurement process itself. This is the measurement problem in a nutshell.

2.3 Some more thoughts on measurement

We said that measurement involves two interacting systems and it is thus natural to represent in our framework the combination of these two systems as the combination of two holistic logics. In fact, we will confine ourselves to

2. A SPECULATIVE LOOK ...

the case of those holistic logics that come from a Hilbert space, i.e. Hilbert space logics.

There is a salient feature of measurement which we call *correlation*. Assume a physicist wants to measure an observable A pertaining to a certain system and assume A can adopt a family of values $(\lambda_i)_{i \in I}$. For this purpose the physicist uses a measuring instrument. He devises an experimental arrangement in which the system to be measured interacts with the measuring instrument in such a way that after measurement the value λ_i assumed by observable A can be read off from a scale or a screen. We may thus view λ_i as the value of an observable pertaining to the measuring instrument. This observable is normally called the pointer observable. So, measuring A means *correlating* it with the pointer observable of the measuring instrument. It is a special convenience of the pointer observable that its values can be read off from a scale or a screen. We have to reflect this notion of *correlation* of observables in our logical framework in the treatment of the measurement problem.

We said that in classical physics measurement doesn't pose a problem. Intuitively, in classical physics measurement is just 'looking' at the system to be measured and the state of that system is not changed in measurement. We will have to reflect this trivial feature of measurement in classical physics. It will have to be reflected as a limiting case similar to the way the logic of classical mechanics appears as a limiting case in the framework of holistic logics as expressed in the limiting case theorem.

To summarise, in our logical treatment we will have to reflect the following characteristics of the process of measurement:

- 1) Combining two systems
- 2) Correlating two systems
- 3) The classical nature of the measuring system (instrument)
- 4) The projection postulate ("collapse of the wave function")
- 5) Classical measurement as a limiting case

2.4 Combining and correlating Hilbert space logics

Given two Hilbert space logics $\mathcal{L}_1$, $\mathcal{L}_2$ with the languages Fml_1 and Fml_2 respectively. We introduce the connective $\otimes$ to form formulas of the 'combined' language. If $\alpha \in Fml_1$ and $\beta \in Fml_2$, then $\alpha \otimes \beta$ is a formula of the combined language. We take these formulas to be the atomic formulas formulas of the combined language. We will use the symbol $\otimes$ both as denoting this connective and the algebraic operation tensor for vectors as well as for the operation of combining consequence relations. A word of caution is in order here. The reader is advised not to think of $\alpha \otimes \beta$ as saying something like 'α and β'. Rather, he may think of $\alpha \otimes \beta$ as making sense only in connection with the combined system. He may, intuitively, think of the combined language as talking about the 'whole', i.e. the combined system, and view the connective $\otimes$ as the 'connective of wholeness'.

DEFINITION 10.5. Given a Hilbert space logic $\mathcal{L}_H$, let $(x_i)_{i \in I}$ be a family of vectors of H containing at most one non-zero vector which, however, may contain the zero vector. Assume $y \in H$ is non-orthogonal to all non-zero members of this family. Then we say that $\vdash_y$ is a superposition of $(\vdash_{x_i})_{i \in I}$.

We will in the sequel, for the sake of brevity, use the term state familiar from quantum mechanics also for denoting consequence relations of a Hilbert space logic.

We may look at this logical concept of a superposition in various ways. First, superpositions may be viewed as encoding all its (consistent) components or 'containing' all the information of its (non-zero) components, since it is non-orthogonal to each of its (non-zero) components. Second, observe that a superposition can be revised so as to yield any of its non-zero components or, put differently, superpositions can, in principle, 'collapse' into each of its non-zero components because a superposition is non-orthogonal to any of its non-zero components.

DEFINITION 10.6. Let $\mathcal{L}_1 = \mathcal{L}_{H_1, \Psi_1}$, and $\mathcal{L}_2 = \mathcal{L}_{H_2, \Psi_2}$ be Hilbert space logics. We define the combination $\mathcal{L} = \mathcal{L}_1 \otimes \mathcal{L}_2 =: \mathcal{L}_{H_1 \otimes H_2}, \Psi$ of $\mathcal{L}_1$ and $\mathcal{L}_2$ as follows. Define Ψ by $\Psi(\alpha \otimes \beta) = \Psi_1(\alpha) \otimes \Psi_2(\beta)$ and as usual for the propositional connectives. Given two consequence relations $\vdash_x$ and $\vdash_y$ of $\mathcal{L}_1$ and $\mathcal{L}_2$ respectively. Then define $\vdash_x \otimes \vdash_y := \vdash_{x \otimes y}$. If one of the component logics is one-dimensional, the combined logic is called a cell.

It is routinely verified that the result of combining Hilbert space logics as described is in fact again a Hilbert space logic. In the sequel we also omit the subscripts.

The following lemma expresses an elementary property of the Hilbert space tensor product.

LEMMA 10.7. *Suppose $\vdash_{x \otimes y} \alpha \otimes \beta$. Then $\vdash_x \alpha$ and $\vdash_y \beta$. Moreover, we have $\vdash_x \otimes \vdash_0 = \vdash_0$, i.e. combining any consequence relation with the inconsistent (zero) consequence relation yields the inconsistent (zero) consequence relation.*

DEFINITION 10.8. Given a Hilbert space logic $\mathcal{L}$ and a family of formulas $A = (\alpha_i)_{i \in I}$ such that for every $i \in I$ $\Psi(\alpha_i)$ is either one-dimensional or the zero-space. Assume that its non-zero members are mutually orthogonal

2. A SPECULATIVE LOOK ...

(one-dimensional subspaces). We assume at least one $\Psi(\alpha_i)$ to be non-zero. Then we say A is an observable of $\mathcal{L}$. We think of α_i, in abuse of notation, as having the form $A = \lambda_i$, the λ_i's being the values of the observable A.

We may for instance think of the index set in the above definition as the reals.

Obviously, the above definition is motivated by the physical concept of an observable represented by a Hermitian operator with non-degenerate eigenvalues. It does not capture the analog e of the fact that the eigenvectors of such an operator form a complete orthonormal system. This is not necessary for our purposes. Eigenvectors are by definition non-zero. Note, however, that in the above definition we allow for the inconsistent consequence relation, i.e. the one corresponding to the zero vector (space), for reasons which will become obvious later. Take for instance the observable energy in the case of the harmonic oscillator. This observable can assume only certain discrete values, namely those of the form $(n + 1/2)\omega$. Nevertheless we may represent this observable as a family of the form $(E = \lambda_i)_{i \in I}$ with I being the set of real numbers. Those real values which are not assumed as values of the energy of the harmonic oscillator then correspond to the illegitimate state, i.e the zero (inconsistent) consequence relation.

We now define the concept of a correlation of observables.

DEFINITION 10.9. Let $\mathcal{L}_1$, $\mathcal{L}_2$ be Hilbert space logics. Let $A = (A = \lambda_i)_{i \in I}$ and $B = (B = \lambda_i)_{i \in I}$ be observables of $\mathcal{L}_1$ and $\mathcal{L}_2$ respectively. Given a consequence relation $\vdash_{x \otimes y}$ of $\mathcal{L}_1 \otimes \mathcal{L}_2$. We say that $\vdash_{x \otimes y}$ is a pure correlation with respect to A and B if $\vdash_{x \otimes y} A = \lambda_i \otimes B = \lambda_i$ for some $i \in I$. We say that $\vdash_y$ is a correlation with respect to A and B if it is either a pure correlation with respect to A and B or a superposition of pure correlations. If a correlation is not pure, we call it an entanglement.

2.5 Representing the measuring instrument

We said that in our treatment of the measurement problem we want to reflect the classical nature of the measuring instrument. The theorem we called the limiting case theorem tells us how classical logic appears as a limiting case in the general framework of holistic logics. By the limiting case theorem for holistic logics the classical limiting case of a holistic logic may be viewed as a complete classical theory or a family of complete classical theories. We saw that this way of passing to the limit can be equally viewed as being in the direction from nonmonotonicity to monotonicity or from non-commutativity to commutativity. The classical limiting case is of the form $\mathcal{L}_\Sigma$ for some complete classical theory Σ or a family of such logics.

We now make the simple connection between holistic logics of the form $\mathcal{L}_\Sigma$, i.e. complete classical theories, with Hilbert space logics. Namely, we can represent a complete classical theory as a one-dimensional Hilbert space logic in a natural way.

So given a logic $\mathcal{L}_\Sigma$ for some complete classical theory Σ. We have $\mathcal{L}_\Sigma = \langle \mathcal{C}, F_\Sigma, \rightarrow \rangle$, where $\mathcal{C}_\Sigma = \{\vdash_\Sigma, 0\}$, $\vdash_\Sigma$ is defined by: $\alpha \vdash_\Sigma \beta$ if $\alpha \rightarrow \beta \in \Sigma$. Now, $\mathcal{L}_\Sigma$ can be presented as a one-dimensional Hilbert space logic as follows. Given a one dimensional Hilbert space. Then the lattice of closed subspaces consists of two elements only, $Sub(H) = \{\langle x \rangle;, \{0\}\}$. Define the function $\Psi : Fml \rightarrow Sub(H)$ by: $\Psi(\alpha) = langlex\rangle$ if $\alpha \in \Sigma$, $\Psi(\alpha) = \{0\}$ else. Then $\mathcal{L}_\Sigma = \mathcal{L}_{H,\Psi}$. We will, from now on, view $\mathcal{L}_\Sigma$ as the one-dimensional Hilbert space logic defined above.

We may think of Σ as the theory of a classical physical system at a certain time t, i.e. the set of statements or formulas true of this system at time t. We may think of such a formula as having the form $A = \lambda$ with the intuitive meaning that the observable A pertaining to a classical system *has* the value λ. In this case we have $(A = \lambda) \in \Sigma$ or, equivalently, this says that $A = \lambda$ is a true statement about this system. Note that for any $\mu \neq \lambda$, $A = \mu$ is not in Σ or, equivalently, is a false statement about the system.

Let us reformulate this as

LEMMA 10.10. *Given a one-dimensional Hilbert space logic $\mathcal{L}_\Sigma$. Then the following conditions are equivalent:*

- *(i) α is true*

- *(ii) $\alpha \in \Sigma$*

- *(iii) $\vdash_x \alpha$*

It is easy to see that the following holds. It reflects our intuition that an observable of a classical system *has* exactly one value.

LEMMA 10.11. *Let $\mathcal{L}$ be a one-dimensional Hilbert space logic and $(A = \lambda_i)_{i \in I}$ an observable of $\mathcal{L}$. Then there exists a unique $j \in I$ such that $A = \lambda_j$ is true in $\mathcal{L}$ and $A = \lambda_i$ is false in $\mathcal{L}$ for $i \neq j$.*

The limiting case results allow us to look at the logics of the form $\mathcal{L}_\Sigma$ as limiting cases in a twofold way. We may say that, given a holistic logic, then the limit of this logic is of the form $\mathcal{L}_\Sigma$, but we may also regard the limits of any of its consequence relations as being of this form. For Hilbert space logics this means that we may either say that the limit of a Hilbert space logic is either a single one-dimensional Hilbert space logic or a family of one-dimensional Hilbert space logics. From the physical point of view the latter version seems preferable since it reflects the fact that phase space is the classical analogue of Hilbert space and phase space may, as explained above, be viewed as a family of complete classical theories.

2. A SPECULATIVE LOOK ...

Let us come back to the crucial question of the classical nature of the measuring instrument. How are we to reflect this?

We do not have much freedom in this. In the light of the above there are two possibilities. Either we treat the measurement instrument as a single one-dimensional Hilbert space logic or a family of one-dimensional Hilbert space logics. We opt for the latter version for reasons which will become obvious later. We call such a family, reminiscent of phase space in classical mechanics, a phase logic. Thus phase logics are families of the form $(\mathcal{L}_j)_{j \in J}$ each member being a one-dimensional Hilbert space logic.

DEFINITION 10.12. A phase logic is any family $(\mathcal{L}_j)_{j \in J}$ of one-dimensional Hilbert space logics.

We now have to define the concept of an observable for phase logics. As before it has to be a family of formulas of a certain kind.

DEFINITION 10.13. Let $\mathcal{L} = (\mathcal{L}_j)_{j \in J}$ be a phase logic. Let $(A = \lambda_i)_{i \in I}$ be a family of formulas. We say that A is an observable of the phase logic $\mathcal{L}$ if it is an observable of all its members.

We have now defined the notion of an observable both for Hilbert space logics and the limiting case of phase logics.

We now have to say how to combine Hilbert space logics with phase logics and how to correlate observables if one of the systems is a phase logic.

DEFINITION 10.14. Let $\mathcal{L}$ be a Hilbert space logic and $(\mathcal{L}_j)_{j \in J}$ a phase logic. We define the combined system to be the family $(\mathcal{L} \otimes \mathcal{L}_i)_{i \in I}$. Call a family of this sort, i.e. a family of cells, a register. Given two observables $(A = \lambda_i)_{i \in I}$ and $(B = \lambda_i)_{i \in I}$ of these systems respectively. We say that a family of states $(\vdash_i)_{i \in I}$, $\vdash_i$ being in cell i, is a correlation of A and B if each member is a correlation of A and B viewed as observables in each cell.

We now define measurement.

DEFINITION 10.15. Let $\mathcal{L}_1$ be a Hilbert space logic and $\mathcal{L}_2 = (\mathcal{L}_j)_{j \in J}$ be a phase logic. Let A and B be observables of $\mathcal{L}_1$ and $\mathcal{L}_2$ respectively. A process of interaction between $\mathcal{L}_1$ and $\mathcal{L}_2$ is called a measurement of A by $\mathcal{L}_2$ with pointer observable B if the state of the combined system 'after interaction' is a non-zero correlation of A and B. The phase logic $\mathcal{L}_2$ is called the measurement instrument, B its pointer observable. We say that $\mathcal{L}_2$ measures observable A of $\mathcal{L}_1$ via the pointer observable B. If $\mathcal{L}_1$ is itself a phase logic, we say that $\mathcal{L}_2$ measures observable A of $\mathcal{L}_1$ via B if it does so for each member of $\mathcal{L}_1$.

The following lemma expresses a simple observation which is, however, crucial to the approach presented here.

LEMMA 10.16. *The correlation with which the process of measurement ends up according to Definition 10.15 is pure, i.e. not an entanglement.*

Proof. We need to see that the correlation is pure in each cell. What do correlations in a cell look like? Generally, a correlation is of the form $A = \lambda_i \otimes B = \lambda_i$ for some $i \in I$ (viewed as a consequence relation) or a superposition of such pure correlations. But now recall that in a one-dimensional Hilbert space logic $B = \lambda_i$ is true for exactly one index ,say j. That is $A = \lambda_i \otimes B = \lambda_i$ is non-zero for for at most the index j and zero for the others. This means that a cell contains at most one non-zero correlation. Then, by the definition of measurement, there exists exactly one non-zero correlation, which then must be pure. ∎

Our definition of measurement reveals, as the reader may easily realise, a shortcoming of the semi-formal treatment of measurement presented here. Namely, in our definition of measurement we used the terms 'interaction' and 'after interaction', which have no counterparts in the formal framework as it is now. We are aware of this. An adequate formal treatment of measurement must be capable of formalising the concept of interaction. It seems that this is not possible in the framework we have at our disposal so far. Intuitively we can, as a makeshift, try to make sense of the above terms, if we think of the formal framework as being implemented somehow, say as a computer system. We may then think of Hilbert space logics for instance not just as abstract concepts but as physically real systems which can (physically) interact.

2.6 Decomposition, projection, revision in measurement

Let us take a closer look at correlations, if we combine an arbitrary Hilbert space logic with a phase logic. This is in our picture the case relevant to measurement. The following is a simple but crucial observation. It is an immediate consequence of the above lemma.

PROPOSITION 10.17. *Let $\mathcal{L}$ be an arbitrary Hilbert space logic and $(\mathcal{L}_j)_{j \in J}$ a phase logic. Recall that the combined system is a register, namely the family of cells $(\mathcal{L} \otimes \mathcal{L}_j)_{j \in J}$. Let $(A = \lambda_i)_{i \in I}$ and $(B = \lambda_i)_{i \in I}$ be observables of the two systems respectively. Suppose $(\mathcal{L}_j)_{j \in J}$ measures observable A via the pointer observable B. Then,'after measurement', the state of cell j has the form $(A = \lambda_{i_j} \otimes B = \lambda_{i_j})$, where i_j is the unique index $i \in I$ for which $B = \lambda_i$ is true in $\mathcal{L}_j$.*

We assume that at any given a certain time t a Hilbert space logic viewed as an implemented systems *is* in a certain state.

We now make the assumption that, given two Hilbert space logics $\mathcal{L}_1$ and

2. A SPECULATIVE LOOK ...

$\mathcal{L}_2$, then the fact that $\mathcal{L}_1 \otimes \mathcal{L}_2$ is in state $x \otimes y$, more precisely $\vdash_x \otimes \vdash_y$ implies that $\mathcal{L}_1$ is in state x and $\mathcal{L}_2$ is in state y.

Let us take a closer look at what happens in the process of measurement as it is represented in the picture painted so far.

So given a Hilbert space logic $\mathcal{L}$ and a phase logic $(\mathcal{L}_j)_{j \in J}$. Let A be the observable to be measured and B the pointer observable.

Assume that before measurement the state of $\mathcal{L}$ is a superposition of 'eigenstates of A', which we informally write as $\sum_i | A = \lambda_i \rangle$. This notation is not Hilbert space notation but just means that the state before measurement is a superposition of 'eigenstates of A'. Now, by our definition of measurement, after measurement each cell $\mathcal{L} \otimes \mathcal{L}_j$ is in a correlation which we, in the same informal notation, write as $\sum_i | A = \lambda_i \rangle \otimes | B = \lambda_i \rangle$. We saw that this correlation is in fact a pure correlation of the form $A = \lambda_k \otimes B = \lambda_k$ for some $k \in I$.

Let us now first consider the case where the measuring phase logic consists of a single one-dimensional Hilbert space logic. The register then consists of one cell. This cell is in a state of the form $A = \lambda_k \otimes B = \lambda_k$ and the system to be measured is in the state $A = \lambda_k$, more precisely in the state corresponding to this proposition. The original state of the system measured, which we assumed to be a superposition of 'eigenstates' of A, was decomposed so to speak and projected onto the 'eigenstate' $A = \lambda_k$. In particular, the state may have changed. We may view the proposition $A = \lambda_k$ and the corresponding state as the result of projection ('collapse of the wave function'). Moreover, we may in our picture of implemented systems, view a register and thus also a cell as a sort of data structure used for storing the result of measurement. If $A = \lambda_k$ is the result of measurement, the 'contents' of the cell is $A = \lambda_k \otimes B = \lambda_k$.

In the case just considered it was clear what the state after measurement is. This is less so in the case where we have a genuine phase logic, i.e. a genuine *family* of one-dimensional Hilbert space logics $(\mathcal{L}_j)_{i \in J}$. In this case it is not obvious what the state after measurement is since we have a proper family of cells $(\mathcal{L} \otimes \mathcal{L}_j)_{j \in J}$ that might 'store' the results of measurement. All we we know in our picture is that cell j is in the state $A = \lambda_{i_j} \otimes B = \lambda_{i_j}$, where i_j is that index i for which $B = \lambda_i$ is true in $\mathcal{L}_j$.

We could assume that exactly one of these states is the unique *actual* state after measurement and one of the values is the real value. Another way of looking at this, which is very much in the general spirit of our picture, is to view the *family* of the values stored in *all* the cells as the outcomes of the measuring process. More precisely, we could view the measurement process as a process of decomposing (disentangling) the superposition $\sum_i (A = \lambda_i)$ (in informal notation) into all its components and storing these component in the register. On this view all the values λ_i would be equally real results of the measurement process. We hit upon a certain parallel here between our logical model and Everett's many world interpretation according to which all values an observable can assume are realised in measurement, in different worlds however. In any case we saw that the state of the composite system after measurement as it is represented in our picture is never an entanglement.

Let us now look at the case of repeated measurement.

So again given a phase logic $\mathcal{L}_j)_{j \in J}$ measuring the observable A of some Hilbert space logic $\mathcal{L}$ via the pointer observable B. We know that whatever we consider the state of $\mathcal{L}$ after measurement, this state is an eigenstate of A, a state corresponding to a proposition of the form $A = \lambda$. We would now like to conclude from what we already know that this state (after measurement) is left unchanged when the measurement is repeated. It seems, however, that this is not possible without an additional assumption concerning the nature of measurement. Namely, the following scenario is compatible with the picture painted so far. Suppose the first measurement ended up in an eigenstate $A = \lambda$. Assume that the repetition of the measurement ends up in a different eigenstate $A = \mu$ with $\mu \neq \lambda$. This scenario is consistent with our finding that the state after measurement is an eigenstate of the observable to be measured but it is at odds with the projection postulate of quantum mechanics. How are we to remedy this? So far we haven't said anything about the relationship between the states before and after measurement. In case the state before measurement is a superposition of eigenstates of A, we found that whatever we consider the state after measurement, this state is obtained by *revising* (in the sense of the theory of Hilbert space logics) the state before measurement by a certain proposition. If we have $A = \lambda$ as the result of measurement, then the corresponding state (after measurement is the result of revising the state before measurement by this proposition. In particular, in case the state before measurement is a superposition of eigenstates of A, then the states before and after measurement are *non-orthogonal*. This is the principle we postulate in order to cope with the idempotence of measurement. The argument is then simply this. We know that the result of measurement is an eigenstate of A. Repeating the measurement yields again an eigenstate. Since different eigenstates are orthogonal we conclude, using the principle postulated above, that the repeated measurement leaves the state unchanged. This is *idempotence* of measurement.

The same argument yields that classical measurement does not involve a change of state. The reason is that in a classical measurement the state before measurement is always an 'eigenstate of the observable to be measured.

2.7 Is the Hilbert space formalism the whole story? Leggett's macrorealism

Let us here remind the reader that our motivation in the last section was purely formal. We asked ourselves the question whether and, if how, the process of measurement can be formalised in the framework of Hilbert space logics. Since in that framework 'classical' and 'quantum' reside side by side with classical being a limiting case, we chose to represent the measuring instrument as classical and the system to be measured as arbitrary, i.e. either classical or genuinely quantum. It is a completely different question, however, whether things are that way in nature. This we don't know. The question is whether the limiting case in our formal framework is realised in nature or not. Is there classical reality to which classical mechanics applies or not? We don't know. But we also know that this is a question on which, at present, there is no universal agreement among physicists.

In fact, in several papers Anthony Leggett put forward his view on physical reality and the formalism of quantum mechanics which has become known as *macrorealism*. Here are some quotations from [37].

"There is simply no convincing evidence that macroscopic superpositions of the type... exist in nature"

"A second reason for reluctance to consider the possibility outlined above lies at a more philosophical level. With few exceptions (who include David Bohm), scientists of the last 300 years or so have been deeply committed to a form of reductionism which holds, in effect, that the behaviour of a complex system of matter must be simply the the sum of the behaviour of its constituent parts."

"In this essay I will try to defend three claims. The first is that the classic quantum measurement paradox, so far from being a non-problem, is a sufficiently glaring indication of the inadequacy of quantum mechanics as a total world view that it should motivate us actively to explore the likely direction in which it will break down."

"Indeed, Bohr, and with greater sophistication Reichenbach were able to to develop an interpretation of the quantum-mechanical formalism which is consistent within its self-imposed limits precisely by postulating a radically different ontological status for microscopic entities such as electrons or neutrons and the macroscopic apparatus which performs the measurement. In the words of a famous quotation by Bohr: 'Atomic systems should not even be thought of as possessing definite properties in the absence of a specific experimental set-up designed to measure these properties."

" The point of view I am proposing - namely that quantum mechanics may not be the whole truth about the physical world- is likely to be strongly antithetical to the views of many ..."

Of course Leggett's macrorealism raises problems. First, if he were right, we would have completely different, even seemingly contradictory descriptions of the microscopic and the macroscopic physical world respectively. But what is macroscopic? What is microscopic? Second, it would mean that, since macroscopic objects may be regarded as aggregates of microscopic objects, classical mechanics should be reducible to quantum mechanics which, however, doesn't seem to be the case.

The question is whether the state of a macroscopic object can be a superposition. Can Schröedinger's cat be "half alive and half dead"? Something like that has never been observed. The explanation given by those believing in the universal validity of the formalism of quantum mechanics is a phenomenon called decoherence. We need not go into this in detail here. It is enough to say this here. Decoherence is a quantum effect arising from the interaction of macroscopic reality with its environment to the effect that superpositions of macroscopic states cannot be observed. So, on this view, even in the macroscopic world superpositions do exist, but, due to an effect called decoherence, they cannot be observed.

Another problem raised by Leggett's macrorealism is the general problem of reductionism. Can the behaviour of macroscopic objects in principle be reduced to that of microscopic objects? If there are no macroscopic superpositions, then the answer is no because, if the formalism of quantum mechanics applies, we never get rid of superpositions. The cat's being 'half alive and half dead' can then not be ruled out as a possible state of the cat.

If Leggett is right, then the Hilbert space formalism is not the whole story. It governs the genuine quantum world but not the classical (macroscopic) world. What then is the whole story? What is the proper mathematical framework describing uniformly both quantum and classical reality?

The ready made answer that the whole story is "quantum mechanics + classical mechanics" rather than just quantum mechanics (Hilbert space formalism) seems to be unsatisfactory since it can hardly be considered a unified framework. This becomes especially apparent when it comes to describing mixed systems containing both classical and quantum components as would be the case of measurement. All we can, from the logical point of view, say here is that, at the logical level, such a unified framework exists in the form of the framework of Hilbert space logics including the limiting case of classical logic represented by one-dimensional Hilbert space logics.

The explanation offered for the apparent absence of superpositions in our macroscopic experience by those who adhere to the view that the formalism of quantum mechanics applies universally is, as already mentioned, a quantum mechanical effect called decoherence. On the other hand it seems that absence of superpositions is a precondition for classical (predicate) logic in which it is taken for granted that objects *possess* definite properties. If the adherents to decoherence were right, would this then mean that the possibility of classical logic is due to a quantum effect? This question is at least worth discussing.

3 Logical monadology?

It is a good habit not to engage in metaphysical speculation in a scientific book, and the authors hope to have abided by this habit despite the fact that quantum mechanics always offers a temptation, if not an invitation, for metaphysical speculation. Now, at the end of the book, we cannot refrain, however, from directing the reader's attention to certain parallels between the findings of this book, which we hope are scientific in nature, and a famous philosophical treatise which has a reputation for being metaphysical in nature. We would like to emphasise that it is solely such parallels, certain formal analogies so to speak, that we think are worth presenting to the reader, and we will refrain from any metaphysical claims whatsoever. We consider these parallels surprising, and perhaps they are not accidental. The metaphysical treatise mentioned is Leibniz's *Monadology*. The *Monadology* is as famous as it is hard to understand. It is a metaphysical treatise dealing with what Leibniz considers the ultimate constituents of the world. At the same time it is a treatise on the relationship between the world as a whole and its parts.

What are, according to Leibniz, the ultimate constituents of the world? According to Leibniz there exist such ultimate constituents which he calls *monads*. The *Monadology* consists of a list of 90 aphorisms each stating a certain properties of the monads. Leibniz makes no attempt to motivate or discuss these properties any further. Let us summarise a few of the properties that Leibniz ascribes to the monads.

Paragraph 3 of the *Monadology* reads: "These monads are the true atoms of nature. In a word, they are the elements of things". And what are the monads? They are 'introduced' in paragraph 1: "Monads, which are our concern here, are nothing other than simple substances." The monads have no parts. In paragraph 3 we read: "Now where there are no parts, neither extension nor shape nor division is possible." The monads are, according to Leibniz, completely self-contained, self-sufficient entities. Paragraph 18 says: "They enjoy self-sufficiency (autarkia) that renders them the source of their internal actions and makes them, so to speak, incorporeal automata."

How then are they related? How can these completely self-sufficient entities be related at all? First, the monads have, due to their self-sufficiency, no direct access to anything outside themselves. Paragraph 7 contains the famous sentence: "The monads have no windows through which anything can come in or go out." It is therefore interesting to hear in paragraph 56 that the monads are, despite their isolated and self-sufficient existence, highly interconnected. The essential concept describing their mutual relationship is that of *mirroring*. The monads mutually *mirror* each other. Every monad is a mirror of any other monad and thus a *mirror* of the whole universe. Paragraph 56 contains the remarkable statement: "Now this interconnection and accommodation of every created thing to every other, of all to each, gives every simple substance relations that express all the other so that each one is a living mirror of the universe."

Now, what are the parallels between the *Monadology* and the formal results of this book? What in particular are the entities in our logical framework that could be viewed as 'corresponding' to the monads of the *Monadology*?

In order to explain this let us imagine a student of mathematics trying to study the *Monadology*. Such a person has high standards of clarity, and it is perfectly possible that Leibniz's famous treatise is not up to these standards. The student could say: "What precisely does Leibniz mean by self-sufficiency, mirroring, no windows? My impression is that this is metaphorical language in a metaphysical treatise. Clarify!" One could then certainly make some attempts at clarification. But it is easily possible that the following situation would arise. The young mathematician could say: "Thank you for trying to help me. But, to be honest, things are still not as clear to me as they should be.

Let me propose the following. Last term I completed a course on algebra dealing with abstract algebraic structures such as groups, rings, fields etc. Such structures are defined by certain abstract axioms. In these axioms the symbols $+$ or $*$ have no concrete meaning at all and these axioms do not say anything about the nature of the elements of the structures in question. It is therefore often helpful for a better understanding of the abstract structures defined by the axioms to have a concrete model. Take for instance group theory. Here we have the abstract axiom system defining groups. This is fine from the purely abstract point of view. But in order to get a feeling for groups it is also necessary to have concrete models of the axioms in which the symbol $+$ for instance has a concrete meaning and in which the elements of the group are well defined objects. What I am saying is that in group theory it is also desirable to consider concrete groups.

Could you perhaps help me in my endeavour to understand the *Monadology* by providing a sort of model in which terms like self-sufficiency, mirroring, no windows have a concrete and precise meaning and in which the objects called monads are well defined objects, perhaps even well defined mathematical objects? Perhaps this could give me a feeling of what

3. LOGICAL MONADOLOGY?

Leibniz has in mind."

In this we could probably help the young mathematician. Namely, we could give him a copy of this book saying: "Read chapters 8 and 9 of this book. There you will find the concept of a holistic logic and in particular that of a Hilbert space logic. Think of the consequence relations of a holistic logic as the monads. To say that consequence relations of a Hilbert space logic are self-sufficient (self-contained) is obviously meaningful. Take our theorem on self-referential soundness and completeness. We may reasonably take this theorem as describing a sort of self-awareness of the states of a holistic logic. Moreover, look at what we called no windows theorems. The no windows theorems essentially say that the set of statements proved by a consequence relation of a holistic logic has no model. If this set of statements were consistent, it would have a model and the consequence relation would thus talk about some external reality. This does not fit into the picture of 'no windows'. What a state of a holistic logic and in particular a Hilbert space logic proves is all about itself, about its own internal working so to speak, and about nothing else. It has perfect introspection but cannot look outside itself.

What is in this model the meaning of mirroring? How does one state of a holistic logic mirror another state? Well, this is what we called encoding. Recall that any state of a holistic logic encodes any state non-orthogonal to it.

As mentioned, Leibniz describes the monads as 'incorporeal automata'. This fits surprisingly well into the picture. In fact, in our model, the activity of the monads is logical deduction, more precisely logical deduction about themselves. And logical deduction is an activity which Leibniz, in anticipation of modern logic, considered automatic or at least automatable. Calculemus!"

Perhaps the student of mathematics with a strong sense of clarity would take these remarks as an incentive to continue reading the *Monadology* with the above in mind. Perhaps he would then find even more parallels between the *Monadology* and holistic logics. But, probably, he would also find out about the limitations of the picture.

BIBLIOGRAPHY

[1] C. Alchourrón, P. Gärdenfors, D. Makinson. On the logic of theory change: Partial meet contractions and their associated revision functions. *Journal of Symbolic Logic* **50**, pp. 185-205, 1985
[2] I.Amemiya, H.Araki. A remark on Piron's paper. *Publications Inst. Math.Sci., Kyoto Univ.*, Ser.A **2**, pp. 423-427, 1966
[3] G. Antoniou. *Nonmonotonic Reasoning*, MIT Press, 1997
[4] G. Birkhoff, J. von Neumann. The logic of quantum mechanics. *Annals of Mathematics* **37**, pp. 823-843, 1936
[5] G. Boolos. *The Logic of Provability*, Cambridge University Press, 1993.
[6] D. Bohm. *Wholeness and the Implicate Order*, Routledge, 1980
[7] M. Born. Quantenmechanik der Stossvorgänge. *Zeitschrift f. Physik*, **38**, pp. 863-867, 1926
[8] M. Born, P. Jordan. Zur Quantenmechanik, *Zeitschrift f. Physik* **34**, pp. 858-888, 1925
[9] M. Born, P. Jordan, W. Heisenberg. Zur Quantenmechanik II, *Zeitschrift f. Physik* **35**, pp. 557-615, 1926
[10] J. Bub. *Interpreting the quantum world*, Cambridge University Press, 1999
[11] C. Cohen-Tannoudji, B. Diu, F. Laloë. *Quantum Mechanics* Vol. 1, 2. John Wiley, 1977
[12] M.L. Dalla Chiara. Quantum Logic. In Gabbay and Guenthner (eds.) *Handbook of Philosophical Logic* Vol. *III*, pp. 427-469, 1986. Revised version in *Handbook of Philosophical Logic*, Second edition, Volume **6**, pp. 129-228, Kluwer, 2001.
[13] M.L. Dalla Chiara. Quantum logic and physical modalities. *J. Philosophical Logic*, **6**, 391-404, 1977
[14] M.L. Dalla Chiara, R. Giuntini, R. Greechie. *Reasoning in Quantum Theory*, Kluwer, 2004
[15] A. Einstein, B. Podolsky, N. Rosen. Can the quantum mechanical description of physical reality be considered complete? *Physical Review* **47**, pp. 777-780, 1935
[16] K. Engesser. Characterisation of classical Hilbert lattices. In P. Hitzler and G. Kalmbach (eds.) *Begabtenförderung im MINT- Bereich*, Band **5**, pp. 1-8, Aegis-Verlag, 2000
[17] K. Engesser, D.M. Gabbay, Quantum logic, Hilbert space, revision theory. *Artificial Intelligence* **136**, pp. 61-100, 2002
[18] R. Feynman. *Lectures on Physics*. Addison Wesley, 2006
[19] U. Friedrichsdorf. *Einführung in die klassische und intensionale Logik*. Vieweg, 1992
[20] D.M. Gabbay. *Investigations in Modal and Tense Logic with Applications to Problems in Philosophy and Linguistics*. Dordrecht, 1976
[21] D.M. Gabbay. *Labelled Deductive Systems*. Clarendon Press, Oxford, 1996
[22] D. M. Gabbay. *Fibring Logics*. Oxford University Press, 1999
[23] D. M. Gabbay. Dynamics of Practical Reasoning: A position paper In *Advances in Modal Logic 2*, K. Segerberg, M. Zakhryaschev, M. de Rijke and H. Wansing, eds. CSLI Publications, CUP, 179-224, 2001.
[24] D.M. Gabbay. Theoretical Foundations for nonmonotonic reasoning in expert systems. In K.R. Apt (ed.) *Proceedings NATO Advanced Study Institute on Logics and Models of Concurrent Systems*, pp. 439-457, Springer-Verlag, Berlin, 1985.
[25] J.-Y. Girard. From Foundations to Ludics. The Bulletin of Symbolic Logic **9**, pp. 131-168, 2003
[26] R.H. Goldblatt. Semantic analysis of orthologic. *J. Philosophical Logic*, **3**, pp. 19-35, 1974
[27] P. Halmos. *Introduction to Hilbert Space an the Theory of Spectral Multiplicity*, 2nd edition, Chelsea, New York, 1957
[28] G.M. Hardegree. The Conditional in Quantum Logic. *Synthese*, **29**, pp. 63-80, 1974
[29] W. Heisenberg. Uber die quantentheoretische Umdeutung kinematischer und mechanischer Beziehungen. *Zeitschrift fur Physik*, **33**, pp. 879-893, 1925
[30] D. Hilbert, P. Bernays. *Grundlagen der Mathematik* II, Springer, Berlin, 1970

[31] S.S. Holland. Orthomodularity in Infinite Dimensions, A Theorem of M. Solèr. *Bulletin of the American Mathematical Society*, **32**, pp. 205-234, 1995
[32] G. Kalmbach. *Orthomodular Lattices. London Math. Soc. Monographs*, Vol. **18** Academic Press, London and New York, 1983
[33] H. Katsuno and K. Sato. A unified view of consequence relation, belief revision and conditional logic. In G. Crocco, L. Farinas del Cerro, and A. Herzig, editors, *Conditionals : From Philosophy to Computer Science*, pp. 33-66, Oxford University Press, 1995.
[34] S. Kochen, E. Specker. The problem of hidden variables in quantum mechanics. *Journal of Mathematics and mechanics* **17**, pp. 59-87, 1967
[35] S. Kraus, D. Lehmann, and M. Magidor. nonmonotonic reasoning, preferential models and cumulative logics. *Artificial Intelligence*, **44**, pp. 167-207, 1990
[36] H. A. Keller. Ein nichtklassischer Hilbertscher Raum. *Mathematische Zeitschrift*, **172**, 41-49, 1980.
[37] A. Leggett. Reflections on the quantum measurement paradox. In *Quantum Implications*, B.J. Hiley, F.D. Peat, eds., Routledge, pp. 85-104, 1987
[38] D. Lehmann. Nonmonotonic Logic and Semantics, *Journal of Logic and Computation*, **11**, pp. 229-256, 2001
[39] D. Lehmann. Connectives in quantum and other cumulative logics. Technical Report Hebrew University of Jerusalem, 2002
[40] D. Lehmann, K. Engesser, D.M. Gabbay. Algebras of measurements: the logical structure of quantum mechanics. *International Journal of Theoretical Physics* **45**, 698-723, 2006
[41] D. Lehmann, M. Magidor. What does a conditional knowledge base entail? *Artificial Intelligence*, **55**, pp. 1-60, 1992
[42] D. Makinson and P. Gärdenfors. Relation between the logic of theory change and nonmonotonic logic. In *The Logic Theory of Change*, A. Fuhrmann and H. Morreau, eds. pp. 185–205. Lecture Notes in AI, **465**, Springer Verlag, 1991.
[43] R. Mayet. Some Characterizations of the Underlying Division Ring of a Hilbert Lattice by Automorphisms. *International Journal of Theoretical Physics*, **37**, pp. 109-114, 1998
[44] P. Mittelstaedt. *Quantum Logic*. D. Reidel Publishing Company, 1978
[45] D. Monk. *Mathematical logic*. Springer, 1976
[46] R.C. Moore. Semantical Considerations on Nonmonotonic Logic. *Artificial Intelligence* **25**, pp. 75-94, 1985
[47] F.D. Peat. *Einstein's Moon*, Contemporary Books, 1990
[48] C. Piron. *Foundations of quantum mechanics*. W.A. Benjamin, 1976
[49] K. Popper. Birkhoff and von Neumann's Interpretation of Quantum Mechanics. *Nature*, **219**, pp. 682-705, 1968
[50] A. Prestel. On Solèr's Characterization of Hilbert spaces. *Manuscripta Math.*, **86**, pp. 225-238, 1995
[51] H. Putnam. Is logic empirical? in R.S. Cohen and M.W. Wartofsky (eds), *Boston Studies in the Philosophy of Science*, Vol **5**, Reidel-Dordrecht, pp. 216-241, 1969,
[52] M. Redéi. *Quantum Logic in Algebraic Approach*. Kluwer Academic Publishers, 1998
[53] M. Redéi (ed.)*John von Neumann: Selected Letters*, vol. **27** of *History of Mathematics*. London Mathematical Society - American Mathematical Society, 2005
[54] W. Rudin. *Real and Complex Analysis*. McGraw Hill, 1974
[55] A. Savile. *Leibniz and the Monadology*. Routledge, 2000
[56] E. Schrödinger. Quantisierung als Eigenwerproblem. *Annalen der Physik* **79**, pp. 1-16, 1926
[57] R.M. Smullyan. *Forever Undecided*. Oxford University Press, 1987
[58] R.M. Smullyan. *Gödel's Incompleteness Theorems*. Oxford University Press, 1992
[59] M.P. Solèr. Characterization of Hilbert spaces with orthomodular spaces. *Communications in Algebra* **23**, pp. 219-234, 1995

[60] J. von Neumann. Mathematische Begründung der Quantenmechanik. *Göttinger Nachrichten*, pp. 1-57, 1927

[61] J. von Neumann. Wahrscheinlichkeitstheoretischer Aufbau der Quantenmechanik. *Göttinger Nachrichten*, pp. 245-272, 1927

[62] J. von Neumann. *Mathematische Grundlagen der Quantenmechanik*, Berlin, Springer, 1932

INDEX

Banach space, 22
Birkhoff-von Neumann, 43
Bohm, 98
Bohr, 94
Boolean algebra, 12
Born, 38

characterisation theorem, 35
classical Hilbert lattice, 33
classical state, 86
closed subspace, 23
completeness theorem, 10
Conjunction M-algebra, 82
consequence relation, 17
consequence revision system, 106
consistent, 5

deduction theorem, 6
deductive system, 3

encodedness, 84, 119
Engesser, 34
Engesser-Gabbay extension, 142
EPR, 96
experimental proposition, 46

Feynman, 151
fibred, 116
finite-dimensional holistic logic, 119
Finkelstein, 75
Fourier expansion, 26
full introspection, 154

Gabbay, 117
Girard, 101
GKLM model, 18

$\mathcal{H}$-model, 115
Hardegree, 33
Heisenberg, 37
Hermitian operator, 39
Hilbert lattice, 33
Hilbert space logic, 133
holistic logic, 118

Implication M-algebra, 81
inference operation, 17
internalising connective, 111

John von Neumann, 38
Jordan, 38

Keller, 34
Kochen-Specker, 53

language of propositional logic, 3
lattice, 11
Leggett, 171
Lehmann, 18
Lehmann extension, 143
Leibniz, 173
limiting case theorem, 89, 128
Lindenbaum algebra, 12
Lindenbaum's lemma, 6
logical consequence, 1

M-algebra, 56
maximal consistent, 5
Mayet, 34
Mittelstaedt, 29
modus ponens, 4
Monadology, 173

no windows theorem, 86, 125
nonmonotonic logic, 15
normed space, 22

orthogonal complement, 23
orthomodular lattice, 28
orthomodular space, 31
orthomodular space logic, 133
orthonormal basis, 24

phase space, 47
Piron, 33
pointer, 83
Popper, 44
projection, 23
proof operator, 141
Putnam, 75

representation theorem, 135

Sasaki hook, 82
Schütte, 53
Schrödinger, 38
self-referential completeness, 103, 123
self-referential soundness, 103, 123
Smullyan, 121
Solèr's theorem, 34
strongly separable M-algebra, 83
symmetry condition, 137

tensor product, 42

uncertainty relations, 16

www.ingramcontent.com/pod-product-compliance
Ingram Content Group UK Ltd.
Pitfield, Milton Keynes, MK11 3LW, UK
UKHW021319180426
11947UKWH00015B/1330